NATIONAL NUCLEAR ENERGY SERIES
Manhattan Project Technical Section

Division III—Volume 1 B

THE THEORY OF ISOTOPE SEPARATION AS APPLIED TO THE LARGE-SCALE PRODUCTION OF U^{235}

THE THEORY OF ISOTOPE SEPARATION AS APPLIED TO THE LARGE-SCALE PRODUCTION OF U^{235}

by

KARL COHEN

Director, Atomic Energy Division,
The H. K. Ferguson Company; formerly Director,
Theoretical Division, SAM Laboratories

Edited by

GEORGE M. MURPHY

Washington Square College, New York University;
formerly at SAM Laboratories, Columbia University

FOREWORD

The wartime project for development of atomic energy was a remarkable feat of cooperation and accomplishment by government, science, industry, labor, and the military services aimed exclusively at the military application of atomic energy. Our present national atomic energy program, expanding upon the previous developments, is directed not only toward the assurance of national security but also toward the realization of the immense potential benefits atomic energy holds for our civilization. The record of progress and the results of extensive scientific investigations and engineering development are contained in the National Nuclear Energy Series. This knowledge, which offers the basis of world-wide benefits from nuclear science, is being published in the established scientific tradition, not solely to meet the precise needs of science but also in support of the high goals of the American people set forth in the Atomic Energy Act. The work reported in this series is a tribute to all the scientists engaged in both the Manhattan Project and the postwar Atomic Energy Commission program.

Gordon Dean, Chairman
U. S. Atomic Energy Commission

ACKNOWLEDGMENT

The Manhattan Project Technical Section of the National Nuclear Energy Series embodies results of work done in the nation's wartime atomic energy program by numerous contractors, including Columbia University. The arrangements for publication of the series volumes were effected by Columbia University, under a contract with the United States Atomic Energy Commission. The Commission, for itself and for the other contractors who contributed to this series, wishes to record here its appreciation of this service of Columbia University in support of the national nuclear energy program.

PREFACE

This volume is one of a series which has been prepared as a record of the research work done under the Manhattan Project and the Atomic Energy Commission. The name Manhattan Project was assigned by the Corps of Engineers, War Department, to the far-flung scientific and engineering activities which had as their objective the utilization of atomic energy for military purposes. In the attainment of this objective, there were many developments in scientific and technical fields which are of general interest. The National Nuclear Energy Series (Manhattan Project Technical Section) is a record of these scientific and technical contributions, as well as of the developments in these fields which are being sponsored by the Atomic Energy Commission.

The declassified portion of the National Nuclear Energy Series, when completed, is expected to consist of some 60 volumes. These will be grouped into eight divisions, as follows:

Division I — Electromagnetic Separation Project
Division II — Gaseous Diffusion Project
Division III — Special Separations Project
Division IV — Plutonium Project
Division V — Los Alamos Project
Division VI — University of Rochester Project
Division VII — Materials Procurement Project
Division VIII — Manhattan Project

Soon after the close of the war the Manhattan Project was able to give its attention to the preparation of a complete record of the research work accomplished under Project contracts. Writing programs were authorized at all laboratories, with the object of obtaining complete coverage of Project results. Each major installation was requested to designate one or more representatives to make up a committee, which was first called the Manhattan Project Editorial Advisory Board, and later, after the sponsorship of the Series was assumed by the Atomic Energy Commission, the Project Editorial Advisory Board. This group made plans to coordinate the writing programs at all the installations and acted as an advisory group in all matters affecting the Project-wide writing program. Its last meeting was held on Feb. 9, 1948, when it recommended the publisher for the Series.

The names of the Board members and of the installations which they represented are as follows:

Atomic Energy Commission	
Public and Technical Information Service	Alberto F. Thompson
Technical Information Division, Oak Ridge Extension	Brewer F. Boardman
Office of New York Operations	Charles Slesser, J. H. Hayner, W. M. Hearon *
Brookhaven National Laboratory	Richard W. Dodson
Carbide & Carbon Chemicals Corporation (K-25)	R. B. Korsmeyer, W. L. Harwell, D. E. Hull, Ezra Staple
Carbide & Carbon Chemicals Corporation (Y-12) †	Russell Baldock
Clinton Laboratories ‡	J. R. Coe
General Electric Company, Hanford	T. W. Hauff
General Electric Company, Knolls Atomic Power Laboratory	John P. Howe
Kellex Corporation	John F. Hogerton, Jerome Simson, M. Benedict
Los Alamos	R. R. Davis, Ralph Carlisle Smith
National Bureau of Standards	C. J. Rodden
Plutonium Project	
Argonne National Laboratory	R. S. Mulliken, H. D. Young
Iowa State College	F. H. Spedding
Medical Group	R. E. Zirkle
SAM Laboratories §	G. M. Murphy
Stone & Webster Engineering Corporation	B. W. Whitehurst
University of California	R. K. Wakerling, A. Guthrie
University of Rochester	D. R. Charles, M. J. Wantman

* Represented Madison Square Area of the Manhattan District.

† The Y-12 plant at Oak Ridge was operated by Tennessee Eastman Corporation until May 4, 1947, at which time operations were taken over by Carbide & Carbon Chemicals Corporation.

‡ Clinton Laboratories was the former name of the Oak Ridge National Laboratory.

§ SAM (Substitute Alloy Materials) was the code name for the laboratories operated by Columbia University in New York under the direction of Dr. H. C. Urey, where much of the experimental work on isotope separation was done. On Feb. 1, 1945, the administration of these laboratories became the responsibility of Carbide & Carbon Chemicals Corporation. Research in progress there was transferred to the K-25 plant at Oak Ridge in June, 1946, and the New York laboratories were then closed.

Many difficulties were encountered in preparing a unified account of Atomic Energy Project work. For example, the Project Editorial Advisory Board was the first committee ever organized with representatives from every major installation of the Atomic Energy Project. Compartmentation for security was so rigorous during the war that it had been considered necessary to allow a certain amount of duplication of effort rather than to permit unrestricted circulation of research information between certain installations. As a result, the writing programs of different installations inevitably overlap markedly in many scientific fields. The Editorial Advisory Board has exerted itself to reduce duplication in so far as possible and to eliminate discrepancies in factual data included in the volumes of the NNES. In particular, unified Project-wide volumes have been prepared on Uranium Chemistry and on the Analysis of Project Materials. Nevertheless, the reader will find many instances of differences in results or conclusions on similar subject matter prepared by different authors. This has not seemed wholly undesirable for several reasons. First of all, such divergencies are not unnatural and stimulate investigation. Second, promptness of publication has seemed more important than the removal of all discrepancies. Finally, many Project scientists completed their contributions some time ago and have become engrossed in other activities so that their time has not been available for a detailed review of their work in relation to similar work done at other installations.

The completion of the various individual volumes of the Series has also been beset with difficulties. Many of the key authors and editors have had important responsibilities in planning the future of atomic energy research. Under these circumstances, the completion of this technical series has been delayed longer than its editors wished. The volumes are being released in their present form in the interest of presenting the material as promptly as possible to those who can make use of it.

The Editorial Advisory Board

The Manhattan Project Technical Section of the National Nuclear Energy Series is intended to be a comprehensive account of the scientific and technical achievements of the United States program for the development of atomic energy. It is not intended to be a detailed documentary record of the making of any inventions that happen to be mentioned in it. Therefore, the dates used in the Series should be regarded as a general temporal frame of reference, rather than as establishing dates of conception of inventions, of their reduction to practice, or of occasions of first use. While a reasonable effort has been made to assign credit fairly in the NNES volumes, this may, in many cases, be given to a group identified by the name of its leader rather than to an individual who was an actual inventor.

COLUMBIA UNIVERSITY PROJECT FOREWORD

Government-supported research on nuclear energy first occurred in early 1940 when certain funds made available by the Army, Navy, and National Bureau of Standards were used for experiments at Columbia University. The subsequent history of this project as it expanded in many other universities and industrial laboratories is told in detail in the Smyth Report. Two of the major programs began at Columbia University: the uranium-graphite pile reactor and the gaseous diffusion method of uranium 235 separation.

Lack of sufficient laboratory space at Columbia for the development of both programs led to the transfer in the spring of 1942 of the uranium-graphite reactor program to the University of Chicago. The program of separation of uranium 235 by gaseous diffusion, and various ancillary projects, developed rapidly until not only all available space in the Columbia University buildings was occupied but, in addition, an even greater amount of rented space.

The project for the separation of uranium isotopes by gaseous diffusion was first carried forward through various government contracts under the direction of Harold C. Urey, Professor of Chemistry, and John R. Dunning, Assistant Professor of Physics. In 1943 the work was unified under a single contract with the Manhattan Engineer District with Professor Urey as Director. An account of the research on the major assignment, the separation of the uranium isotopes, is to be found elsewhere in the National Nuclear Energy Series. The volumes that this preface introduces deal with the general mathematical theory of isotope separation, with some new experimental methods of separating isotopes, with spectroscopic properties of uranium compounds, and with the chemical and physical properties of heavy water and other deuterium compounds. Most of this work was done at Columbia, but parts of it were carried on at the National Bureau of Standards and at The Johns Hopkins University.

George B. Pegram
Chairman, Columbia University Division
of War Research, 1941-1945

PUBLISHER'S NOTE

Although every effort has been made to ensure accuracy in references, at the time of publication of this book some of the other volumes of the Series had not been completed. It is therefore possible that some of the references to other volumes are in error. It is hoped that the extensive cross checking which has been done in the preparation of this volume has resulted in keeping such errors to a minimum.

AUTHOR'S PREFACE

In principle the subject matter of this volume is the reports issued by the theoretical division of the SAM Laboratories and its antecedents, which the author was privileged to lead. The work was done over a period of years, from 1940 to 1945, under contract to various government agencies. However, we have in some instances drawn on outside sources for special topics, and some of the material in the volume is new. Since the reports have not been and probably will not be published, we have avoided distracting the reader with references in the body of the text which would be meaningless to him. At the end of each chapter a reference section of relevant reports is appended for purposes of record. Specific references to published articles are footnoted as they occur in the text.

Although the theory was developed piecemeal in response to various demands, it is exposed as a logical whole according to our present understanding and experience. In Chaps. 1 to 5, on various aspects of cascade theory, certain generalities that are common to many different separation methods are collected. Chapter 6 treats the centrifugal method. Chapter 7 is composed of sections briefly outlining two-phase separations (including chemical exchange and distillation) and the thermal-diffusion method, and a section presenting some remarks on the separation of deuterium.

The studies given are a selection from a much larger volume of material. More emphasis is put on general principles and concepts than on an exhaustive collection of particular results, no matter how elegant. It is thought unlikely that any detailed problem will recur in exactly the same form and that a thorough grasp of the fundamentals will be more useful to the reader. In any event, the most important applications of the theory cannot yet be discussed.

While the theory itself is intact, the reader may find that the motivations and explanations are occasionally somewhat sketchy. This is a result of security deletions. We did not feel that they were important enough to warrant extensive rewriting.

There is no reference to direct experimental proofs of the theory, although they are not lacking. This is only partly a question of space and security. In a large measure the theory is self-evident, in the fashion of demonstrations in thermodynamics.

Karl Cohen

August 1951

ACKNOWLEDGMENTS

Development of the theory in this volume was influenced by many discussions with people whose names appear on no theoretical report. It is a pleasure to acknowledge indebtedness to Drs. J. R. Dunning, E. T. Booth, and H. A. Boorse, all of Columbia University, and Dr. F. G. Slack, of Vanderbilt University, for ideas and criticisms.

A British team, composed of Drs. F. Simon, R. Peierls, K. Fuchs, and N. Kurti, worked out independently the theory of separation as it applied to a particular separation. They obtained substantially the same results, in a different chronological order, corresponding to the different order in which the problems were attacked. Adequate liaison was accomplished in early 1942, at which time a very stimulating comparison of results and viewpoints was made. This and subsequent contacts contributed toward clarity of thought and expression.

Much of the credit for the success of these investigations is due to Dr. H. C. Urey: on an administrative plane, for his unwavering support of the theoretical group; and on a scientific plane, for imparting to us his fine intuitive grasp of the essentials of separation processes. The concept of separative work, which is basic for the comparison of different methods, was first used by him in 1939.

The author is personally indebted to E. V. Murphree, of the Standard Oil Development Company, for allowing him the time to write this volume.

The role of our colleagues in the theoretical division does not really need to be mentioned here; their contributions will be amply demonstrated by the reference sections. Above all, Dr. I. Kaplan, H. Mayer, and Dr. R. D. Present have contributed substantially to the success of this work.

The elegant derivation of the value functions and the separative power in Chap. 1, Sec. 3, is due to Prof. P. A. M. Dirac. The value function itself, Eq. 1.43, was first determined by Prof. R. Peierls. The relaxation-time phenomena for square cascades were worked out with the constant aid of Dr. I. Kaplan. The method of determining the fundamental constants in the steady state is due to Dr. W. I. Thompson, of the Standard Oil Development Company. The mathematical method employed in Chap. 4 on losses was shown to the author by Dr. R. P. Feynman, then at Princeton University. The solution of the fluctuation problem, which avoids solving the differential equation for the second order, was devised by Dr. K. Fuchs.

Without the aid of Mrs. L. W. Goodhart, of the Standard Oil Development Company, in editing and in all phases of preparation of the manuscript, it could never have been finished; special thanks are due to her for this important share of the work.

Karl Cohen

CONTENTS

Page

Foreword v
Preface vii
Columbia University Project Foreword xi
Author's Preface xiii
Acknowledgments xv
Introduction 1

CHAPTER 1

Ideal Cascades 5

CHAPTER 2

Square Cascades 30

CHAPTER 3

Equilibrium Time of a Square Cascade 39

CHAPTER 4

Determination of Cascade Constants 62

CHAPTER 5

The Control Problem 79

CHAPTER 6

Centrifuges 103

CHAPTER 7

Other Separation Methods 126

Appendix A — Roots of a Transcendental Equation 139

Appendix B — Equilibrium Time of Square Cascades for N ~ 1 144

Appendix C — The Holdup Function 147

Appendix D — Rayleigh Distillation 150

Appendix E — Properties of Concurrent Two-phase Elements . . 154

Index 163

INTRODUCTION

The separation of isotopes may be accomplished in many ways. In this volume consideration will be given, in more or less detail, to centrifuges, electromagnetic methods, electrolysis, chemical exchange, thermal diffusion, and distillation. Within each of these categories there is considerable variety. For example, chemical exchange may take place between a liquid phase, between a liquid phase and a vapor phase by countercurrent scrubbing, or between two liquid phases with concurrent contacting.

All these methods, except the mass-spectrograph method in isolated instances, have the common feature that one elementary process does not produce the desired large change in isotopic abundance. The problem of efficiently multiplying elementary processes to reach large end results is thus of utmost practical importance, and indeed of greater importance the smaller the change produced by the elementary process.

Fortunately, although the methods differ both physically and mathematically in fundamental ways, they are sufficiently alike so that their multiplication problems are very closely related. It will therefore be useful to examine thoroughly the multiplication problem per se before taking up any of the several separation schemes in detail.

The purpose of the first five chapters is to develop the theory of cascades as generally as possible in order to permit the widest possible application to future separation problems. Existing applications of the theory will be found in the subsequent chapters on particular separation methods. The point of view taken is that of a development engineer, and the main concern is the significance of the results in terms of cascade design. For example, the effect of losses is investigated in terms of the problem "How much larger do we have to build our stages to get the same quality and quantity of yield for a given loss?" rather than "In a given cascade how will the yield vary with loss?" Consideration is seldom given to problems of the second kind—problems that occur after the cascade is built—although naturally the same equations govern both questions.

In Secs. 1 to 4 of Chap. 1 the concept of an ideal cascade is examined from three radically different points of view; Section 5 introduces the idea of equilibrium time. Section 6 gives the general partial differential equation showing the dependence of mole fraction in a cascade on time and stage number.

The first portion of Chap. 2 is intended to show the steady-state behavior of single countercurrent columns, or square cascades. In Secs. 2 and 3 of the chapter, cascades of countercurrent columns are considered. Section 4 covers the squaring off of the top of a cascade.

Time-dependent phenomena in square cascades are dealt with in Chap. 3, along with the methods for experimental determination of the fundamental constants of a separating element.

Chapter 4 considers the increase in size of a cascade with losses, the efficient combination of plants using different separation processes, and the prime operating problem of the control of a cascade.

Chapter 5 deals with the mathematical theory of the effect of disturbances in operation on the production of the cascade. Chapter 6 describes the different types of centrifuges and gives the partial differential equation of the centrifuge. Other separation methods are presented in Chap. 7.

Only for certain kinds of processes—"infinitesimal" processes, countercurrent columns, and the mass spectrograph—is the theory developed explicitly, in such a way as to permit immediate application. In the last chapter, which includes the deuterium problem, are introduced cascade problems that are not treated in Chaps. 1 to 5. But no one who has absorbed the first five chapters will be at a loss to solve the problems. Likewise, among the infinite variety of nonideal cascades, only square cascades have been selected as being of universal interest. The reader will be able to work out for himself such variants as cascades in the form of a lozenge (the "cascade-of-cascades" in British terminology).

The generality of the treatment still leaves something to be desired, since the entire treatment is restricted to binary isotope mixtures. This is a consequence of the practical origin of the theory.

In a general way an attempt has been made to avoid lengthy mathematical demonstrations, except when they are indispensable to the argument. For this reason there is only a very general treatment of the control problem, which is beset with many purely mathematical difficulties. A detailed study of the equilibrium-time problem of an ideal cascade has also been omitted; here again the problem has more mathematical than physical interest. Some of the sections that are mostly mathematical might perhaps be omitted on a first reading.

They are Secs. 2 and 3 of Chap. 2; Secs. 1 and 2 of Chap. 3; Secs. 1, 2, and 3 of Chap. 5; and Appendixes A and B. In the Appendixes are grouped a number of extensions and amplifications that have considerable practical interest. Appendixes D and E in particular contain special results upon which some work in the later chapters is based.

The many graphs which are included were designed for use as well as illustration, and particular pains were taken to ensure their accuracy. The same applies to the tables, which were in some cases recalculated entirely for this book.

Principles of Notation and Terminology. Because of the wide variety of topics that will be covered in this review, it will not be possible to set up a system of notation that will be general enough to cover every case and at the same time describe each special case naturally. However, certain principles of notation will be followed.

A mole fraction of a particular isotope will always be denoted by the letter N. In our case of binary mixtures, the two mole fractions are of course N and (1−N). In two-phase systems, all lower-case letters will refer to one phase, upper-case letters to the other. Thus N and n will refer to mole fractions in different phases. With every binary mole fraction N we will associate the corresponding molecular abundance ratio R, where

$$R = \frac{N}{1-N} \qquad N = \frac{R}{1+R}$$

Thus corresponding to n there will be $r = n/(1-n)$. As material flows through a separating element, its mole fractions are changed. The mole fractions at the various orifices of a separating element will be distinguished by primes (for example, see Fig. 1.1). Quantities of material—generally expressed in moles per second—are designated by L's.

The term "cascade" is used to mean any connected arrangement of elements. The "elements" are the separative units—individual centrifuges or fractionating columns. A "stage" of a cascade consists of all units operating in parallel on material of the same mole fraction. (In the special case of one element per stage, the terms "element" and "stage" will be used interchangeably.) The feed material of a cascade, generally of natural abundance ratio, is denoted by the subscript zero (N_0); the amount of feed is F moles/sec. The "stage number" (counting the bottom stage as zero) is denoted by s. The mole fraction of material fed into the sth stage is N_s, and the amount is L_s. The subscript S indicates the top stage of a rectifying cascade. The mole fraction at the product end of a cascade is N_P; the amount of product is P. The mole fraction of the waste is N_W; the amount of waste is W.

Other general terminology and notation will be covered in the subsequent introductory discussion of cascades. Equations, figures, and tables are numbered decimally (viz., Eq. 6.4); the first digit indicates the chapter, the number after the decimal the order in the chapter.

Chapter 1

IDEAL CASCADES

Consider a separating element that divides an entering stream of L moles/sec of a bicomponent mixture, with the mole fraction of one component designated as N, into two streams of θL moles and mole fraction N′, and $(1-\theta)$L moles and mole fraction N″, respectively (Fig. 1.1). Of course $0 < \theta < 1$, where θ is known as the "cut." Further suppose that the action of the unit is such that

$$R' = \alpha R \qquad (\alpha > 1) \tag{1.1}$$

The term α is called the "simple-process factor" of the separation.[1]

From the conservation of matter

$$N = \theta N' + (1 - \theta)N'' \tag{1.2}$$

Now suppose that a cascade is built of such elements, the size of each stage being graduated so that the flow entering the *s*th stage is L_s. For purposes of this discussion it is irrelevant whether each stage is built of a single element, whose size differs from stage to stage, or of identical small elements in parallel, whose number varies. The elements are connected as shown in Fig. 1.2, so that the enriched fraction from each element is fed into the stage next above and the other fraction is fed into the stage immediately below it. At the top of the cascade, material is withdrawn continuously at a rate P moles/sec. It should be observed that there is considerable latitude

[1] This term is usually reserved for true equilibrium ratios of abundance ratios (cf. Eq. 6.7 and Eq. 8 of Appendix D) which, however, have no universal relationship to the α of Eq. 1.1. If necessary to distinguish, α of Eq. 1.1 is referred to as the "effective" simple-process factor. True equilibrium factors are denoted by α_0.

about the choice of the flows, L_s. For the instant they are completely arbitrary.

The problem now at hand is one of choosing flows L_s and cuts θ_s to get the most efficient cascade. Clearly, the remixing of materials of different mole fractions must be avoided. That is, it is imperative that

$$R'_{s-1} = R_s = R''_{s+1} \tag{1.3}$$

From Eqs. 1.1 and 1.3, $R_s = R'_{s-1} = \alpha R_{s-1}$, whence

$$R_s = \alpha^s R_0 \tag{1.4}$$

follows immediately by induction. Likewise

$$R''_s = \frac{1}{\alpha} R_s \tag{1.5}$$

From Eqs. 1.2 and 1.5,

$$\theta_s = \frac{1 + \alpha R_s}{(\alpha + 1)(1 + R_s)} \tag{1.6}$$

Considering that section of the cascade above an imaginary line drawn between the sth and (s + 1)th stages (see Fig. 1.2), we must have balance between the amount of materials entering and leaving this whole section, i.e.,

$$\theta_s L_s - (1 - \theta_{s+1})L_{s+1} = P \tag{1.7}$$

$$\theta_s L_s N'_s - (1 - \theta_{s+1})L_{s+1} N''_{s+1} = PN'_S = PN_P \tag{1.8}$$

Substituting Eq. 1.7 in Eq. 1.8,

$$\theta_s L_s (N'_s - N''_{s+1}) = P(N_P - N''_{s+1}) \tag{1.9}$$

Equation 1.9, which is a consequence of the conservation of matter, holds generally for all cascades, even if mixing takes place. Introducing Eq. 1.3, the condition of no mixing, in Eq. 1.9,

$$L_s = \frac{P(N_P - N_s)}{\theta_s (N'_s - N_s)} \tag{1.10}$$

Using the properties of the separating element itself and eliminating N'_s by Eq. 1.1 and θ_s by Eq. 1.6,

$$L_s = \frac{(\alpha + 1)P(N_P - N_s)}{(\alpha - 1)N_s(1 - N_s)} \tag{1.11}$$

Equation 1.11 gives the flow required in the sth stage of a cascade producing P moles/sec of material with mole fraction N_P, and Eq. 1.6 gives the required cut to ensure that no mixing takes place in the

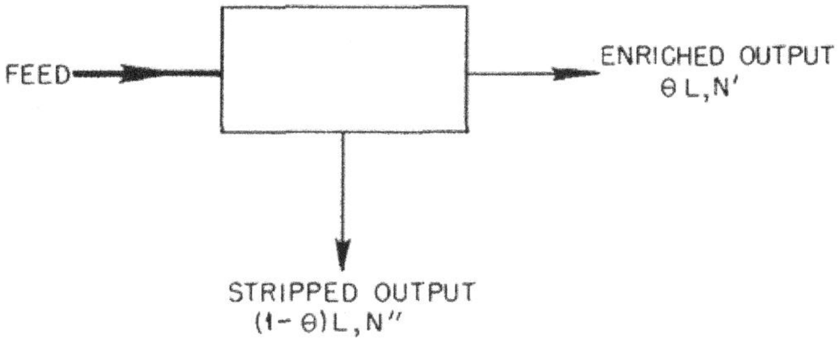

Fig. 1.1—A simple separating element.

cascade. Such a cascade is called an "ideal" or "no-mixing" cascade. In exactly what sense this is the most efficient cascade will be made clear later.

Equation 1.11 can be written as

$$L_s = \frac{\alpha + 1}{\alpha - 1} PN_P \left[1 - \alpha^{s-S-1} + \frac{1}{R_0}(\alpha^{-s} - \alpha^{-S-1})\right] \tag{1.12}$$

by replacing the N's by R's and remembering that $R_s = \alpha^s R_0$ and that $R_P = \alpha^{S+1} R_0$. In this form the L_s may be summed to get the total flow in an ideal cascade. Noting that

$$\sum_{s=0}^{S} \alpha^s = \frac{1 - \alpha^{S+1}}{1 - \alpha}$$

it readily follows that

$$\sum_{s=0}^{S} L_s = \frac{\alpha + 1}{\alpha - 1} PN_P \left[(S + 1)\left(1 - \frac{\alpha^{-S-1}}{R_0}\right) + \frac{\alpha^{-S-1} - 1 + (\alpha^{S+1} - 1)\alpha^{-S}}{R_0(\alpha - 1)}\right]$$

$$= \frac{\alpha + 1}{\alpha - 1} P\left[\frac{(2N_P - 1)\ln R_P/R_0}{\ln \alpha} + \frac{\alpha - (\alpha + 1)N_0}{\alpha - 1} \frac{N_P - N_0}{N_0(1 - N_0)}\right] \tag{1.13}$$

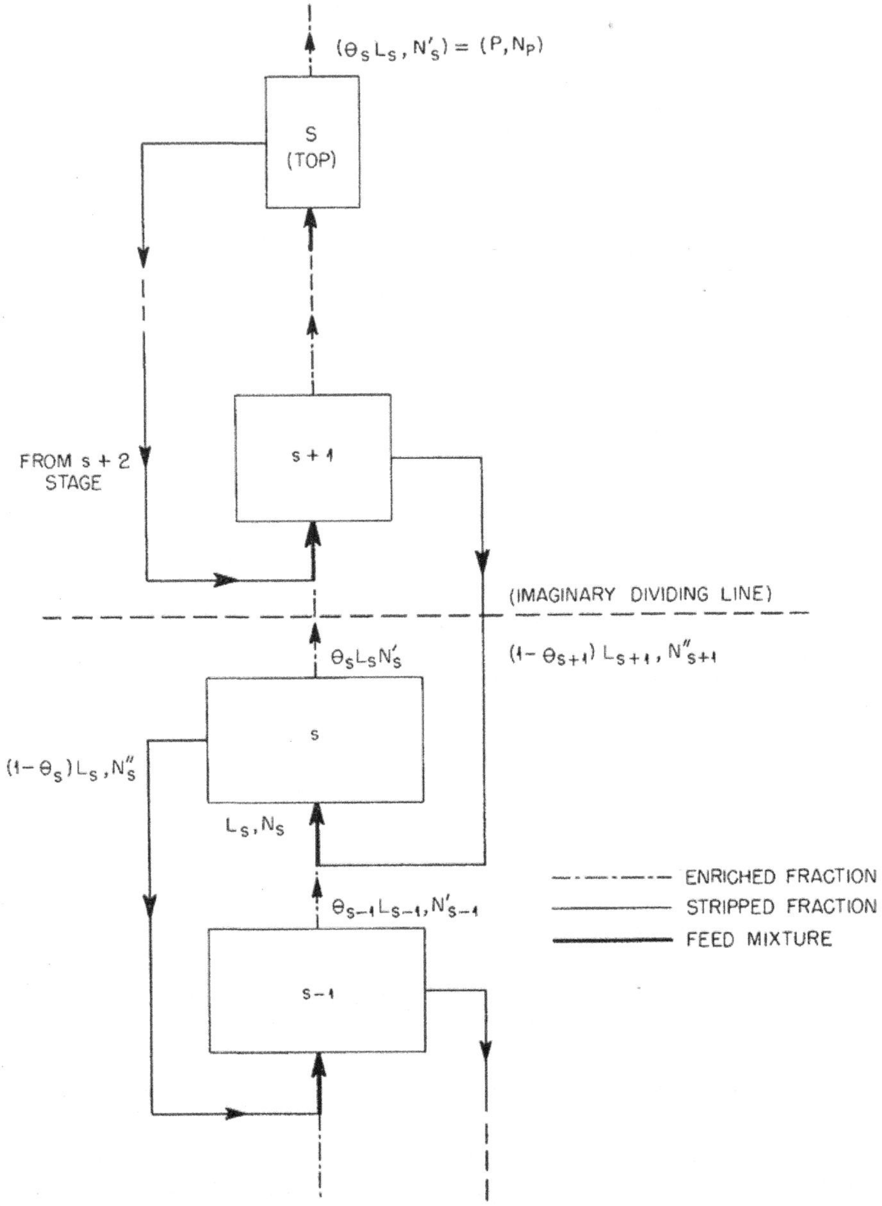

Fig. 1.2—A cascade of simple elements.

Since the size, or number of elements, of a stage is proportional to the flow into the stage, the total flow measures the magnitude of the whole cascade.

1. THE INFINITESIMAL CASE

Because of the special importance of cascades when the simple-process factor is very nearly 1, the limiting forms of the equations just found will be written for

$$\alpha - 1 = \epsilon \ll 1 \tag{1.14}$$

Equation 1.1 can be expressed as

$$N' - N = \frac{R'}{1 + R'} - \frac{R}{1 + R} = \frac{R' - R}{(1 + R')(1 + R)} = (\alpha - 1)N(1 - N') \tag{1.15}$$

which in the limit is the same as

$$N' - N = \epsilon N(1 - N) \tag{1.16}$$

In the no-mixing case Eq. 1.15 becomes

$$N_{S+1} - N_S = \epsilon N_S (1 - N_S) \tag{1.17}$$

and Eqs. 1.4, 1.6, and 1.11 are replaced by the following:

$$R_S = e^{\epsilon S} R_0 \tag{1.18}$$

$$\theta_S = \frac{1}{2} + \frac{\epsilon}{4}(2N_S - 1) \tag{1.19}$$

$$L_S = \frac{2}{\epsilon} \frac{P(N_P - N_S)}{N_S(1 - N_S)} \tag{1.20}$$

The total flow in the cascade may be determined from Eq. 1.20 as follows:

$$\sum_{s=0}^{S} L_S - \int_0^S L_S \, ds = \int_{N_0}^{N_P} L_S \frac{ds}{dN} \, dN \tag{1.21}$$

Now Eq. 1.17 may be written

$$\frac{dN_S}{ds} = \epsilon N_S (1 - N_S) \tag{1.22}$$

Inserting in Eq. 1.21 and dropping the now superfluous subscript s,

$$\sum_{s=0}^{s} L_s = \frac{2P}{\epsilon^2}\int_{N_0}^{N_P} \frac{N_P - N}{N(1-N)}\,\frac{dN}{N(1-N)}$$

$$= \frac{2P}{\epsilon^2}\left[(2N_P - 1)\ln\frac{R_P}{R_0} + \frac{1-2N_0}{1-N_0}\,\frac{N_P - N_0}{N_0}\right] \qquad (1.23)$$

which could have also been obtained directly from Eq. 1.13.

Equation 1.19 shows that in the infinitesimal case θ_s varies very slightly with N_s. Even if—as they must be—ϵ and θ are related by the physical nature of the separating element, ϵ will be constant throughout the cascade, to terms of the order of ϵ^2, and the preceding analysis holds. But for the finite case, if an internal relation between α and θ exists which differs from Eq. 1.6, α will vary from stage to stage, and the equations break down. Practically, this means that the cascade equations will not be universally applicable unless $(\alpha - 1)$ is small.[1]

2. RELATION BETWEEN IDEAL AND NONIDEAL CASCADES

Perhaps the best way of understanding the real significance of ideal cascades is to consider their relation to nonideal cascades. To do this the infinitesimal case must again be treated, this time allowing mixing.

From Eqs. 1.2 and 1.16,

$$N_s'' = N_s' - \frac{\epsilon}{1-\theta_s} N_s' (1 - N_s') \qquad (1.24)$$

using N_s' instead of N_s on the right to the same accuracy as Eq. 1.16. From Eqs. 1.7 and 1.8,

$$(1 - \theta_{s+1})L_{s+1}(N_s' - N_{s+1}'') = P(N_P - N_s') \qquad (1.25)$$

Combining Eqs. 1.24 and 1.25,

$$N_s' - N_{s+1}' + \frac{\epsilon}{1-\theta_s} N_s'(1 - N_s') = \frac{P}{(1-\theta_{s+1})L_{s+1}}(N_P - N_s') \qquad (1.26)$$

Now, from Eq. 1.7, if P is much smaller than L_s and if the values of L_s vary slightly from stage to stage, all the values of θ_s will be

[1] Cf. Appendix D: Rayleigh Distillation.

very close to ½. Passing over to the limit, which is clearly legitimate since N_s varies slowly,

$$\frac{dN}{ds} = 2\epsilon N(1 - N) - \frac{2P}{L}(N_P - N) \tag{1.27}$$

Equation 1.27 gives the rate of increase of N per stage as a function of the flow per stage, here considered as an arbitrary function. Now consider a function such as the total flow in the cascade

$$\text{Total flow} = \int_0^S L\ ds = \int_{N_0}^{N_P} L(N)\frac{ds}{dN}\,dN \tag{1.28}$$

or the total number of elements, each of which is capable of processing G moles/sec,

$$\text{Number of elements} = \frac{1}{G}\int_{N_0}^{N_P} L(N)\frac{ds}{dN}\,dN \tag{1.29}$$

or more generally, the integral of any property proportional to the flow per stage

$$I = \int \phi(N) L\ ds = \int_{N_a}^{N_b} \frac{\phi(N)}{\epsilon N(1 - N)}\,\frac{L^2}{(2L - L^*)}\,dN \tag{1.30}$$

where N_a and N_b are arbitrary; $\phi(N)$ does not change sign in the range $N_a < N < N_b$, and

$$L^* = \frac{2P(N_P - N)}{\epsilon N(1 - N)}$$

the ideal flow of Eq. 1.20. The asterisk (*) is used in this volume to denote the Laplace transform of any function.

The design problem for cascades may then be expressed as follows: Of all possible cascades that produce P moles of product with a mole fraction N_P, starting from raw material of mole fraction N_0, which cascade makes I a minimum? Or, since each function L(N) completely determines a cascade, what function L(N) makes I stationary?

Clearly, I will be a minimum when

$$\frac{d}{dL}\,\frac{L^2}{2L - L^*} = 0 \tag{1.31}$$

that is, when

$$L = L^* \tag{1.32}$$

Therefore the ideal cascade minimizes the number of separating elements, and related quantities, in any section of a cascade.

When $L = L^*$, from Eq. 1.27,

$$\frac{dN}{ds} = \epsilon N(1 - N) \tag{1.33}$$

is half the maximum concentration gradient (which occurs when $P/L = 0$). Optimum rate of production of a stage is that rate which reduces the concentration gradient to half its value at no production.

It is well to bear in mind that the concept of an ideal cascade is associated with the form of I. For example, to minimize expressions such as

$$\text{Total number of stages} = \int_0^S ds = \int_{N_0}^{N_P} \frac{L}{2L - L^*} \frac{dN}{\epsilon N(1 - N)} \tag{1.34}$$

requires $L = \infty$ and not $L = L^*$. In practice it will be necessary to strike a balance between reducing the number of separating elements and reducing the number of stages. Furthermore an ideal cascade requires each stage to be of a different size. For these reasons actual cascades will deviate somewhat from ideal cascades. Nevertheless, the properties of ideal cascades are of great practical importance because of the proposition that follows.

Developing I about $L = L^*$,

$$I = \int \frac{\phi(N)}{\epsilon N(1 - N)} L^* \, dN + \int \frac{\phi(N)}{\epsilon N(1 - N)} \frac{(L - L^*)^2}{L^*} \, dN + \ldots \tag{1.35}$$

The first term on the right is called I^*, the value of I for the ideal case. If $|L - L^*| \leq yL^*$, for $y < 1$, Eq. 1.35 gives, to higher powers of y,

$$|I - I^*| \leq y^2 I^* \tag{1.36}$$

Expressed in words: If the flow in a cascade at no point differs from that in the corresponding ideal cascade by more than 20 per cent, integrals of the type in Eq. 1.30 will not differ from the ideal integrals by more than 4 per cent. It is this stationary property of integrals for the ideal cascade which makes them useful for all cascades.

3. VALUE FUNCTIONS AND SEPARATIVE POWER

Equation 1.23, for the total flow in an ideal cascade, is the ratio of a term

$$P\left[(2N_P - 1)\ln\frac{R_P}{R_0} + \frac{1-2N_0}{1-N_0}\,\frac{N_P - N_0}{N_0}\right]$$

which depends only on the amount and mole fraction of material produced, and a term $\epsilon^2/2$, which is characteristic of the separating element. This suggests that each term separately has physical significance and also that Eq. 1.23 might be established in a more general way.

An attempt should now be made to derive a function U that represents the value of a quantity of separated material, i.e., a function proportional to the number of separating elements required to produce it. Obviously $U = P\,V(N)$, where P is the number of moles of material. Suppose G moles of material is passed through a separating element of the type shown in Fig. 1.1. There results a net change in value

$$\delta U = \theta G\,V(N') + (1-\theta)G\,V(N'') - G\,V(N) \tag{1.37}$$

Taking $(\alpha - 1) = \epsilon \ll 1$ and expanding $V(N')$ and $V(N'')$ in Taylor's series about N,

$$\delta U = V(N)\,[\theta G + (1-\theta)G - G] + \frac{dV(N)}{dN}\left\{\theta G(N'-N) + (1-\theta)G\,[(N''-N)]\right\} + \frac{d^2V(N)}{dN^2}\left[\theta G\,\frac{(N'-N)^2}{2} + (1-\theta)G\,\frac{(N''-N)^2}{2}\right] + \ldots \tag{1.38}$$

The coefficients of $V(N)$ and $dV(N)/dN$ vanish by the conservation of matter. From Eq. 1.2,

$$\theta(N'-N) = -(1-\theta)(N''-N) \tag{1.39}$$

and from Eq. 1.16, $N' - N = N(1-N)$. Thus Eq. 1.38 becomes

$$\delta U = \frac{\theta}{1-\theta}\,\frac{G\epsilon^2}{2}\,\frac{d^2V(N)}{dN^2}\,[N(1-N)]^2 \tag{1.40}$$

Now, specifying that this change in value be independent of N, the following equation must hold:

$$\frac{d^2V(N)}{dN^2} = \frac{1}{[N(1-N)]^2} \tag{1.41}$$

Then Eq. 1.40 becomes

$$\delta U = \frac{\theta}{1-\theta}\,\frac{G\epsilon^2}{2} \tag{1.42}$$

The term δU is called the "separative power" of the element.[1]

The differential expression of Eq. 1.41 defines V(N) except for two integration constants. Following Dirac, these may be chosen by requiring that $V(N_0) = dV(N_0)/dN = 0$, where N_0 is the natural abundance. The first condition is obvious; the second makes V(N) a minimum at $N = N_0$ and so ensures a positive value for V over the entire range of concentrations.

Thus it is found that

$$V = (2N-1)\ln\frac{R}{R_0} + (N-N_0)\frac{(1-2N_0)}{N_0(1-N_0)} \tag{1.43}$$

V will be called the "value function." Obviously it is the value per mole.

Since one element produces a change in value δU, regardless of the mole fraction of the material on which it operates, the number of elements required to produce a change in value ΔU is

$$\text{Total number of elements} = \Delta U/\delta U \tag{1.44}$$

and

$$\text{Total flow} = G \times (\text{number of units}) = G\,\frac{\Delta U}{\delta U} = \frac{2\Delta U}{\epsilon^2}\,\frac{1-\theta}{\theta} \tag{1.45}$$

For the cascade previously considered

$$\Delta U = U_P + U_W - U_0$$

where the subscripts refer to the product, waste, and feed points. With the convention chosen, $U_0 = U_W = 0$ and Eq. 1.45 is merely

[1] Note that ϵ must be a function of θ, as otherwise $\delta U \rightarrow \infty$ as $\theta \rightarrow 1$.

$$\text{Total flow} = \frac{GU_P}{\delta U} = \frac{2P\,V(N_P)}{\epsilon^2}\,\frac{1-\theta}{\theta} \qquad (1.46)$$

It should be pointed out that, in order for Eq. 1.44 to hold, there must be no process taking place in the cascade which changes U except the separation process; that is, there must be no mixing.[1] Equation 1.46 therefore gives the total flow in an ideal cascade, as found previously for the special case[2] $\theta = \frac{1}{2}$.

A different choice of the integration constants is of course permissible. Taking $V(0.50) = dV(0.50)/dN = 0$, which means taking the equimolar mixture instead of the natural abundance as the fiducial point, gives the simple formula

$$U' = PV' = P(2N - 1)\ln R \qquad (1.47)$$

for the value, which has the obvious advantages of depending on only one mole fraction and of being the same for all isotopes. However, an equation like Eq. 1.46 takes the form

$$\text{Total flow} = \frac{G}{\delta U}\,(U'_P + U'_W - U'_0)$$

and since U' is always used in such differences, with two relations between the parameters (Eqs. 1.61 and 1.62), its numerical value is less illuminating.

From the standpoint of calculations, if there is one particular isotope-separation problem, with N_0 known, values of V as defined by Eq. 1.43 are most useful. These are found in Fig. 1.3 for the uranium problem, $N_0 = 1/140$. For a survey of many different separation problems, values of V' are required. They are found in Table 1.1 and Fig. 1.4. V' will be called the "elementary value function," to distinguish it from V, which is the value function relative to natural abundance. U' will be called the "elementary value," and U will be defined as the "value of a quantity of material."

The value functions and the separative power have been derived for infinitesimal processes. They may also be extended to cover the case of large values of α.

[1] Mixing reverses the process in Eq. 1.37 and hence decreases U.

[2] Since θ has not been restricted, in general a more intricate cascade than the one shown in Fig. 1.2 will be required. For example, if $\theta = \frac{2}{5}$, the enriched fraction from stage s will have to be fed into stage s + 3, while the depleted fraction is fed into stage s − 2.

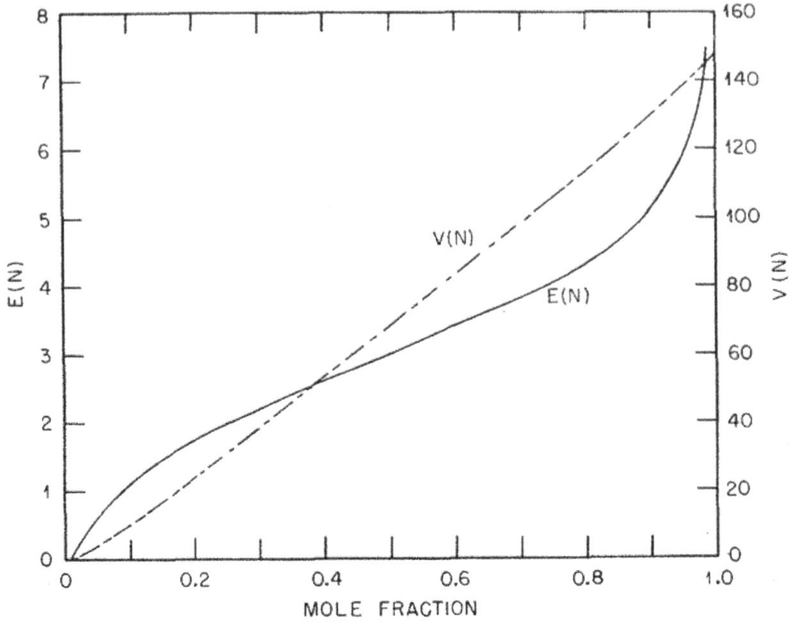

Fig. 1.3—Functions associated with infinitesimal cascade. $N_0 = 0.007143$.

Consider a symmetric process[1] where (cf. Eqs. 1.5 and 1.6)

$$R' = \alpha R \qquad R'' = \frac{1}{\alpha} R \tag{1.48}$$

$$\theta = \frac{1 + \alpha R}{(\alpha + 1)(1 + R)} \qquad 1 - \theta = \frac{\alpha + R}{(\alpha + 1)(1 + R)} \tag{1.49}$$

Then corresponding to Eq. 1.37

$$\frac{\delta U}{G} = \frac{1 + \alpha R}{(1 + R)(\alpha + 1)} V(\alpha R) + \frac{\alpha(1 + R/\alpha)}{(1 + R)(\alpha + 1)} V\frac{R}{\alpha} - V(R) = \text{constant} \tag{1.50}$$

Now introduce s by the equation $R = R_0\alpha^s$ and set

$$(1 + R)\,V(R) = g(R) = g(R_0\alpha^s) = F(s) \tag{1.51}$$

[1] The problem is also soluble for asymmetric processes, where quite complicated functions are found.

Table 1.1—The Elementary Value Function

For $N > 0.50$ use $V'(N) = V'(1 - N)$

$$V' = (2N - 1) \ln \frac{N}{1 - N}$$

N	V'	N	V'
0.001	6.893	0.24	0.5994
0.002	6.188	0.25	0.5494
0.003	5.547	0.26	0.5020
0.004	5.473	0.27	0.4575
0.005	5.240	0.28	0.4155
0.01	4.503	0.29	0.3766
0.015	4.006	0.30	0.3389
0.02	3.736	0.31	0.3040
0.025	3.480	0.32	0.2714
0.03	3.268	0.33	0.2408
0.035	3.085	0.34	0.2122
0.04	2.924	0.35	0.1857
0.045	2.780	0.36	0.1611
0.05	2.650	0.37	0.1384
0.055	2.531	0.38	0.1175
0.06	2.421	0.39	0.09843
0.065	2.320	0.40	0.08108
0.07	2.225	0.405	0.07307
0.075	2.135	0.41	0.06552
0.08	2.052	0.415	0.05836
0.085	1.972	0.42	0.05165
0.09	1.897	0.425	0.04535
0.095	1.826	0.43	0.03945
0.10	1.758	0.435	0.03400
0.11	1.631	0.44	0.02894
0.12	1.514	0.445	0.02430
0.13	1.407	0.45	0.02007
0.14	1.307	0.455	0.01635
0.15	1.214	0.46	0.01283
0.16	1.128	0.465	0.009814
0.17	1.047	0.47	0.007206
0.18	0.9705	0.475	0.005005
0.19	0.8989	0.48	0.003201
0.20	0.8318	0.485	0.001802
0.21	0.7685	0.49	0.000800
0.22	0.7087	0.495	0.000200
0.23	0.6525	0.50	0.000000

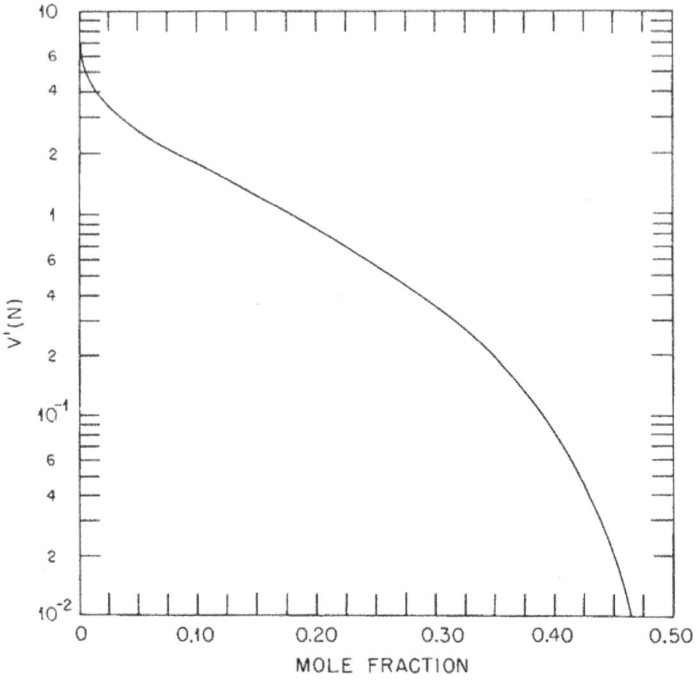

Fig. 1.4—The elementary value function. For values of $N > 0.50$, use $V'(N) = V'(1 - N)$; for $N > 0.46$, $V' = 2(2N - 1)^2$.

With these definitions Eq. 1.50 becomes

$$F(s + 1) + \alpha F(s - 1) - (\alpha + 1)\,F(s) = C(\alpha + 1)\,(1 + R_0\alpha^s) \qquad (1.52)$$

The general solution of Eq. 1.52 is

$$F(s) = C\,\frac{\alpha + 1}{\alpha - 1}\,s(R_0\alpha^s - 1) + A\alpha^s + B \qquad (1.53)$$

where A and B are constants, which gives

$$V(R) = C\,\frac{\alpha + 1}{\alpha - 1}\,\frac{2N - 1}{\ln \alpha}\,\ln\frac{R}{R_0} + \frac{AN}{R_0} + B(1 - N) \qquad (1.54)$$

Setting $V(R_0) = 0$, and establishing the convention that the waste material from the feed stage, of abundance ratio R_0/α, has zero value also, the following relation holds:

$$V(R) = C\frac{\alpha+1}{\alpha-1}\left[\frac{2N-1}{\ln\alpha}\ln\frac{R}{R_0} + \frac{\alpha - N_0(\alpha+1)}{\alpha-1}\;\frac{(N-N_0)}{N_0(1-N_0)}\right] \tag{1.55}$$

Now take the constant $C = \delta U/G$ so that

$$\delta U = G\frac{(\alpha-1)\ln\alpha}{\alpha+1} \tag{1.56}$$

Thus

$$U'' = P\left[(2N-1)\ln\frac{R}{R_0} + \frac{\ln\alpha}{\alpha-1}\;\frac{\alpha-(\alpha+1)N_0}{N_0(1-N_0)}(N-N_0)\right] \tag{1.57}$$

Equations 1.56 and 1.57 are closely analogous to Eqs. 1.42 and 1.43. The variation of U'' with α exhibited in the second term is caused by the varying concentration of by-product waste material and has no theoretical significance. The elementary value for large α is of course still U'.

The separative power varies as $(\alpha-1)^2$ for small $(\alpha-1)$, but for large α increases only as ln. Since the number of elements in a cascade varies inversely as the separative power of an element, it is important to design and operate the elements so that their separative power (Eq. 1.42 or Eq. 1.56) is as large as possible. This problem will be considered in detail in subsequent discussions on particular separation schemes.

It is also possible to calculate the separative power for some more general types of separating elements. If it is assumed that instead of two streams leaving the apparatus of Fig. 1.1, there are many, each of magnitude L^k, and furthermore

$$N^k - N = \epsilon_k N(1-N) \qquad \epsilon_k \ll 1$$

it follows that

$$\begin{aligned}\delta U &= \sum_k L^k\;\; V(N^k) - L\,V(N)\\ &= \sum (L^k - L)\,V(N) + \sum L^k (N^k - N)\frac{dV(N)}{dN}\\ &\qquad + \sum_k L^k\,\frac{\epsilon_k^2}{2}\,[N(1-N)]^2\,\frac{d^2V(N)}{dN^2}\end{aligned} \tag{1.58}$$

The first two terms on the right vanish by the conservation of matter. Hence

$$\delta U = \sum_k L^k \frac{\epsilon_k^2}{2} \tag{1.59}$$

of which Eq. 1.42 is a special case.

For infinitesimal separating elements of the countercurrent type, δU may be evaluated as follows: Consider two large reservoirs (see Fig. 1.5) containing, respectively, M moles of material in which the

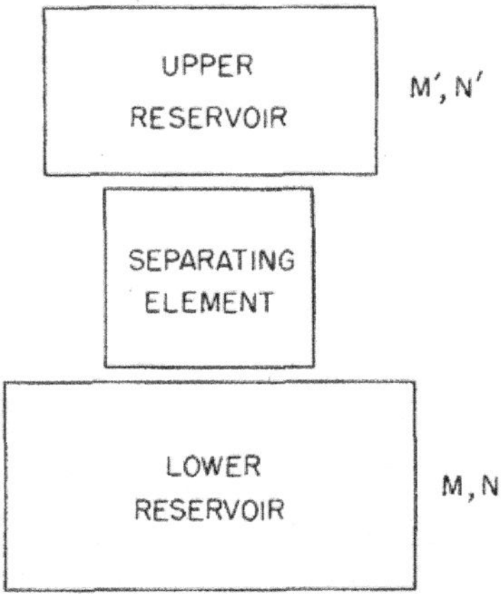

Fig. 1.5—Separative work of a countercurrent element.

mole fraction of desired material is N, and M′ moles of slightly different mole fraction $N' > N$. The separating element puts P moles of mixed isotopes containing τ moles of (pure) desired isotope into the upper reservoir, withdrawing the same amounts from the lower reservoir. The upper reservoir increases in mole fraction by an amount

$$\frac{N'M' + \tau}{M' + P} - N' = \frac{\tau - N'P}{M' + P}$$

and the lower decreases by the amount $(\tau - PN)/(M - P)$. Then

$$\delta U = \delta[M' \, V(N')] + \delta[M \, V(N)]$$

$$= (\tau - NP)\left[\frac{dV(N')}{dN'} - \frac{dV(N)}{dN}\right] = (\tau - NP)\,(N' - N)\,\frac{d^2V(N)}{dN^2}$$

Usually both $\tau - NP$ and $N' - N$ are proportional to $N(1 - N)$; if so

$$\delta U = \frac{(\tau - NP)(N' - N)}{N^2(1 - N)^2} \tag{1.60}$$

The quantity τ, i.e., the amount of desired isotope moved forward, is called the "desired material transport." The factor $\tau - NP$ is called the "net transport." Equation 1.60 is an exceedingly important formula for the theory of countercurrent processes.

The separative powers for two other important special types of elements are fully discussed in Appendix D (Rayleigh Distillation) and Appendix E (Properties of Concurrent Two-phase Elements).

4. CASCADES WITH STRIPPING SECTIONS

In addition to a rectifying section, it will usually be desirable to have a stripping section in the cascade.[1] The function of the stripping section is to economize raw material by recovering the desired isotope from the reject stream of the rectifier. Figure 1.6 shows the relations between rectifier and stripper.

For any cascade, ideal or otherwise, provided there is no loss of material anywhere in the system, the following relations hold:

$$F = P + W \tag{1.61}$$

$$N_0 F = N_P P + N_W W \tag{1.62}$$

where F, P, and W refer to the amount of feed, product, and waste, respectively, and N_0, N_P, and N_W are the corresponding mole fractions.

Taking the mole fractions as given and eliminating first F and then W between Eqs. 1.61 and 1.62, it is found that

$$W = P \frac{N_P - N_0}{N_0 - N_W} \tag{1.63}$$

[1] The terms "rectifier" and "stripper" are taken from fractionating-tower usage. In a tower for producing alcohol from wine, for example, the section above the feed point rectifies the alcohol, i.e., concentrates it, while the lower section strips the waste water, i.e., recovers the last traces of alcohol from the reject.

$$F = P\,\frac{N_P - N_W}{N_0 - N_W} \tag{1.64}$$

which relate the size of the waste stream and the feed stream to the rate of production.

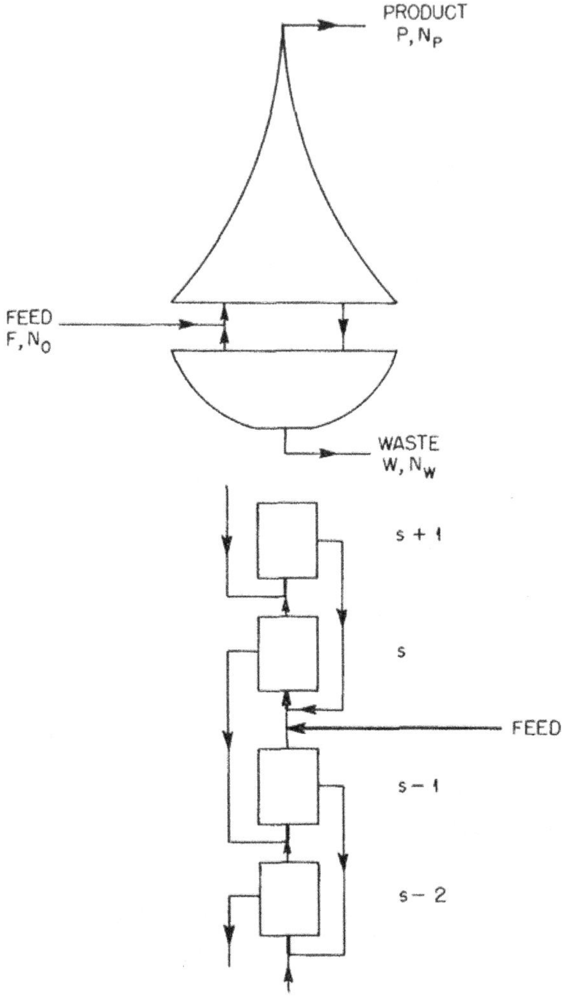

Fig. 1.6—Above, connections between rectifier and stripper. Below, detail of connections.

Since a stripping section is essentially no different from a rectifying section, except that it produces material at a different rate and concentration, the equations for an ideal stripper should be expected to resemble closely those for an ideal rectifier. A moment's consideration, referring again to Fig. 1.2, shows that Eqs. 1.4, 1.9, and 1.11 also hold for the stripping stages of a cascade, with $-W$ and N_W replacing P and N_P, s running from -1 to $-B$ instead of from 0 to S, and $R_W = R_0/\alpha^{b+1}$. Equation 1.11 for the flow in the sth stage of a rectifier becomes

$$L_s = \frac{\alpha + 1}{\alpha - 1} \frac{W(N_s - N_W)}{N_s(1 - N_s)} \tag{1.65}$$

Equation 1.13 becomes

$$\begin{aligned}\text{Total flow (stripper)} &= \sum_{s=-1}^{B} \frac{\alpha + 1}{\alpha - 1} WN_W \left[(\alpha^{B+s+1} - 1) + \frac{1}{R_0}(\alpha^{B+1} - \alpha^{-s})\right] \\ &= \frac{\alpha + 1}{\alpha - 1} W \left[\frac{1 - 2N_W}{\ln \alpha} \ln \frac{R_0}{R_W} \right. \\ &\qquad \left. - \frac{\alpha - N_0(\alpha + 1)}{(\alpha - 1)} \frac{N_0 - N_W}{N_0(1 - N_0)}\right] \\ &= \frac{\alpha + 1}{(\alpha - 1) \ln \alpha} U''_W \end{aligned} \tag{1.66}$$

where U''_W is the value defined by Eq. 1.57. This expression vanishes for $R_W = R_0/\alpha$, in accordance with our convention of counting stage 0 as part of the rectifier.

The total flow in a cascade composed of both rectifier and stripper, combining Eqs. 1.13, 1.64, and 1.66, takes the simple forms

$$\begin{aligned}\text{Total flow (rectifier and stripper)} &= \frac{\alpha + 1}{(\alpha - 1) \ln \alpha} \left[P(2N_P - 1) \ln \frac{R_P}{R_0} \right. \\ &\qquad \left. + W(2N_W - 1) \ln \frac{R_W}{R_0}\right] \\ &= \frac{\alpha + 1}{(\alpha - 1) \ln \alpha} (U_P + U_W) \\ &= \frac{\alpha + 1}{(\alpha - 1) \ln \alpha} (U'_P + U'_W - U'_0) \end{aligned} \tag{1.67}$$

The terms $U_P + U_W$ and $U'_P + U'_W - U'_0$ are of course $\Delta U = \Delta U''$ and $\Delta U'$, respectively. Note that in Eq. 1.67 the distinction between U″ and U has disappeared, since the terms that differ because of different conventions have canceled out.

For the infinitesimal case, in place of Eq. 1.65,

$$L_s = \frac{2}{\epsilon} \frac{W(N_s - N_W)}{N_s(1 - N_s)} \tag{1.68}$$

In place of Eq. 1.66

$$\text{Total flow (stripper)} = \int_{N_0}^{N_W} \frac{2}{\epsilon^2} \frac{W(N - N_W)}{N(1 - N)} \frac{dN}{N(1 - N)} = \frac{2}{\epsilon^2} U_W \tag{1.69}$$

and in place of Eq. 1.67

$$\begin{aligned} \text{Total flow (rectifier and stripper)} &= \frac{2}{\epsilon^2}(U_P + U_W) \\ &= \frac{2}{\epsilon^2}(U'_P + U'_W - U'_0) \\ &= \frac{2}{\epsilon^2}[P\ V(N_P) + W\ V(N_W)] \\ &= \frac{2}{\epsilon^2}[P\ V'(N_P) + W\ V'(N_W) \\ &\qquad - F\ V'(N_0)] \end{aligned} \tag{1.70}$$

For completeness the formulas for the number of stages may be added.

$$\text{Number of rectifying stages} = S + 1 = \frac{\ln R_P/R_0}{\ln \alpha} \tag{1.71}$$

$$\text{Number of stripping stages} = B = \frac{\ln R_0/R_W}{\ln \alpha} - 1 \tag{1.72}$$

$$\text{Number of stages (rectifier and stripper)} = \frac{\ln R_P/R_W}{\ln \alpha} - 1 \tag{1.73}$$

For the infinitesimal case, $\ln \alpha$ is merely replaced by ϵ.

5. EQUILIBRIUM TIME

Obviously some cascades for separating isotopes are very large and contain considerable material in process at concentrations ranging from that of the product to that of the feed. Before the plant can produce concentrated material at the top it must build up these inventories of partly enriched material. The time spent in this process is known as the "equilibrium time" or "relaxation time" of the cascade.

As will be shown later the operation of a cascade from starting-up time might be as follows: At zero time the mole fraction at the top of the cascade is N_0; hence operations begin with no withdrawal. The concentrations increase gradually over the entire cascade, the top of the cascade leading. The product end reaches design mole fraction first, while the rest of the cascade is still below design. Withdrawal at a slow rate is then begun, maintaining the top mole fraction at N_P and gradually increasing to design production as mole fractions in the rest of the cascade build up to their steady-state values (Fig. 1.7). A general definition of the equilibrium time that has meaning for this type of operation or any other is the number of days production that is lost between starting-up time and steady-state production.

A useful approximate expression for the equilibrium time of an ideal rectifier by simple considerations will now be developed.

The total amount of material in process in a section of a cascade is called the "holdup" of the section.[1] In most cascades the amount of holdup in a stage is proportional to the flow in the stage. This is certainly true if the cascade is built up of identical small elements, so that successive stages consist of different numbers of elements in parallel. It also holds when the stages are similar single units scaled to size. The holdup per unit flow will be denoted by h, which is also the average process time per stage, expressed in seconds.

The "net desired material holdup" (N.D.M.H.) is the amount of desired isotope in a going plant over and above that which would be contained in a like quantity of material at normal abundance. It is, for the infinitesimal case,

$$\text{N.D.M.H.} = \int_{N_0}^{N_P} hL_s\,(N_s - N_0)\,ds$$

$$= \frac{2hP}{\epsilon^2}\left[(N_P - 2N_P N_0 + N_0)\ln\frac{R_P}{R_0} - 2(N_P - N_0)\right] \qquad (1.74)$$

The equilibrium time in any case is equal to the N.D.M.H. divided by a suitable average transport of desired isotope over the initial period of operation. Without for the moment specifying too closely the oper-

[1] Nomenclature again borrowed from fractionating-tower terminology.

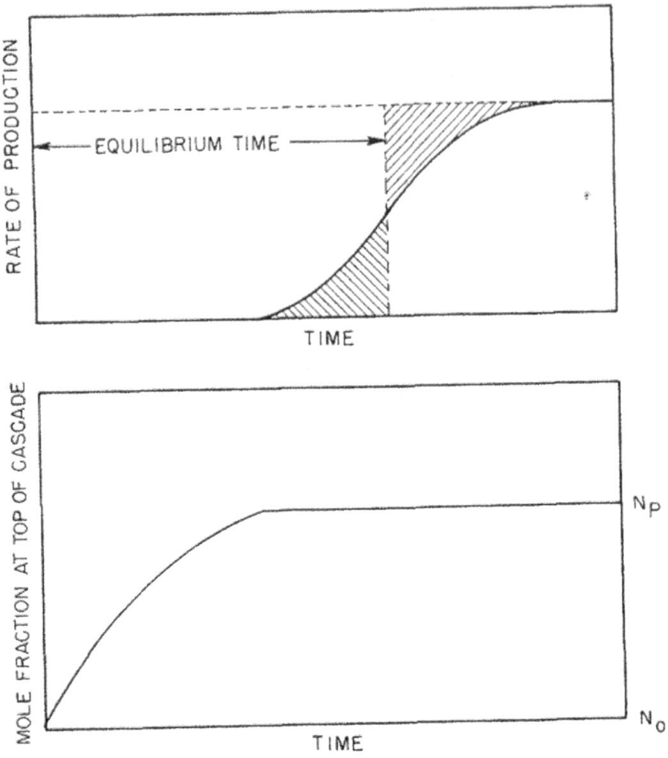

Fig. 1.7—Production as a function of time.

ating schedule, this average transport may be taken as the net transport of the rectifier in the steady state, that is, $P(N_P - N_0)$. This gives

$$\text{Equilibrium time} = \frac{\frac{2h}{\epsilon^2}\left[(N_P - 2N_PN_0 + N_0)\ln\frac{R_P}{R_0} - 2(N_P - N_0)\right]}{(N_P - N_0)}$$

$$\equiv \frac{2h}{\epsilon^2} E(N_P) \qquad (1.75)$$

It should be remarked that the formula for the equilibrium time in the form

$$\text{Equilibrium time} = \frac{\text{N.D.M.H.}}{\text{average net transport}} \qquad (1.76)$$

may be applied to other than ideal cascades. The usefulness of the expression will depend on the appropriateness of the choice of the average net transport, which will not in general be $P(N_P - N_0)$.

The N.D.M.H. is one of the functions that is a minimum for ideal cascades, as is the equilibrium time function $E(N_P)$ derived from it. $E(N_P)$ is independent of P.

6. FUNDAMENTAL EQUATIONS OF ISOTOPE SEPARATION

The partial differential equations describing the time-dependent effects in any infinitesimal cascade are easily derived by a slight extension of the previous analysis.[1] L_s and θ_s are taken as known functions that may vary with time. The mole fraction N_s is also a function of time: $N_s = N_s(t)$.

Referring again to Fig. 1.2, Eq. 1.7 takes the more general form

$$\theta_s L_s - (1 - \theta_{s+1})L_{s+1} = T_s(t) \tag{1.77}$$

where $T_s(t)$ is the amount of (mixed) material transported into the cascade above the sth stage. Explicitly, if there are no leaks or losses,

$$T_s(t) = P + \sum_{j=s+1}^{S} \frac{\partial H_j}{\partial t} \tag{1.78}$$

H_j is the holdup in the jth stage and is also a given function of time. $T_s(t)$ is defined in terms of known functions (by Eq. 1.77 or 1.78) and therefore is known.

Likewise Eq. 1.8, the equation of conservation of desired material, holds in the form

$$\theta_s L_s N_s'(t) - (1 - \theta_{s+1})L_{s+1}N_{s+1}''(t) = \tau_s(t) \tag{1.79}$$

where $\tau_s(t)$ is the amount of desired material transported into the section of the cascade above the sth stage. Unlike T, τ is unknown. Explicitly

$$\tau_s(t) = P\,N_S'(t) + \sum_{j=s+1}^{S} \frac{\partial}{\partial t} H_j N_j'(t) \tag{1.80}$$

[1] For finite α it is difficult to formulate the problem without specifying the separation method.

In using $N_j'(t)$ on the right the difference between the mean concentration of the stage and N_s' has been tacitly neglected. Combining Eqs. 1.77 and 1.79

$$(1 - \theta_{s+1})L_{s+1}\,[N_s'(t) - N_{s+1}''(t)] = \tau_s(t) - N_s'(t)\,T_s(t)$$

Assuming Eq. 1.24 to hold unchanged (what is involved is neglecting the holdup of one stage compared to $\sum_{s+1}^{s} H_j$) and introducing Eq. 1.24 in the last equation,

$$(1 - \theta_{s+1})L_{s+1}\left[N_s'(t) - N_{s+1}'(t) + \frac{\epsilon}{1 - \theta_{s+1}}\,N_s'(1 - N_s')\right] = \tau_s(t) - N_s'(t)\,T_s(t)$$

On passing over to the limit of continuous functions, $\theta_s L_s \rightarrow \frac{1}{2}L(s,t)$, $N_s'(t) \rightarrow N(s,t)$, etc., this becomes

$$\tfrac{1}{2}L(s,t)\left\{-\frac{\partial}{\partial s}N(s,t) + 2\epsilon\,N(s,t)\,[1 - N(s,t)]\right\}$$
$$= \tau(s,t) - N(s,t)\,T(s,t) \qquad (1.81)$$

which is the required generalization of Eq. 1.27. From Eqs. 1.78 and 1.80 it is found[1] that

$$\frac{\partial}{\partial s}T(s,t) = -\frac{\partial}{\partial t}H(s,t)$$
$$\frac{\partial}{\partial s}\tau(s,t) = -\frac{\partial}{\partial t}[H(s,t)\,N(s,t)] \qquad (1.82)$$

Differentiating Eq. 1.81 with respect to s and using Eq. 1.82, $\partial\tau/\partial s$ may be eliminated and a second-order partial differential equation for N will result. Thus Eqs. 1.81 and 1.82, with appropriate boundary conditions, completely determine N.

[1] The integral form of Eq. 1.78, for example, is

$$T(s,t) = P + \int_{s}^{S}\frac{\partial}{\partial t}H(j,t)\,dj$$

and differentiating with respect to s gives Eq. 1.82.

Because special cases of Eqs. 1.81 and 1.82 describe the properties of apparatus in widely differing separation methods and because all the theory so far developed for infinitesimal cascades can be deduced from them by purely formal operations, Eqs. 1.81 and 1.82 are called the "fundamental equations of isotope separation."

If L is not a function of t, and H = hL, then T = P, and Eqs. 1.81 and 1.82 combine to give

$$hL \frac{\partial N}{\partial t} = \frac{\partial}{\partial s}\left[\frac{L}{2}\frac{\partial N}{\partial s} - \epsilon LN(1-N) - PN\right] \tag{1.83}$$

Equation 1.83, with the initial condition $N(s,0) = N_0$, describes the approach to equilibrium of a cascade. For an ideal cascade, L is the extremely complicated function defined by Eqs. 1.18 and 1.20. Equation 1.83 will not be solved for this case, since the utility of the solution is slight. In Chap. 3 it will be solved for the practical (and considerably easier) case of a square cascade, taking L as a constant.

Should L (and also H and ϵ) be a function of t, fluctuating about some mean value, then Eqs. 1.81 and 1.82 describe the effect of fluctuations on separation, which is covered in Chap. 5, on the control problem.

REFERENCES

Cohen, K., Report A-60, October 1941; Report A-96, Jan. 5, 1942.

Cohen, K., Columbia Ser. No. 4L-X28, Dec. 22, 1942 (letter to J. R. Dunning).

Cohen, K., and I. Kaplan, Report A-98, Jan. 14, 1942.

Dirac, P. A. M., British MS, 1941.

Fuchs, K., British MS 29, Dec. 18, 1941.

Fuchs, K., and R. Peierls, British MS 12A, February 1942; British MS 47, June 26, 1942.

Kurti, N., and F. Simon, British MS, 1940.

Peierls, R., British MS 12, summer 1940; British MS 13, September 1940.

Chapter 2

SQUARE CASCADES

Ideal cascades are characterized by a smoothly varying flow function which is next to impossible to realize in practice. Fortunately, as the discussion in Chap. 1 of the relation between ideal and nonideal cascades has shown, it is by no means necessary to adhere strictly to the shape of the ideal cascade.

The outlines of this cascade roughly follow those of an ideal cascade, and the properties of the blocked-off cascade closely resemble those of the ideal cascade. A cascade thus obtained from an ideal cascade is called a "squared-off cascade." Each section of a squared-off cascade is itself a square cascade.

For a square cascade Eq. 1.83 takes the form

$$c_6 \frac{\partial N}{\partial t} = c_5 \frac{\partial^2 N}{\partial s^2} - \frac{\partial}{\partial s}[PN + c_1 N(1 - N)] \tag{2.1}$$

where c_1, P, c_5, and c_6 are constants. Specifically

$$\begin{aligned} c_6 &= hL = \text{holdup per stage} \\ c_5 &= L/2 \\ c_1 &= \epsilon L \end{aligned} \tag{2.2}$$

Proof will be given later that precisely the same equation governs the behavior of a countercurrent centrifuge, a fractionating column, a thermal-diffusion column, and most other countercurrent devices. In each of these cases c_1 and c_5 have different but analogous meanings, and s is replaced by z, the coordinate along the length of the column. A countercurrent unit is therefore equivalent to a square cascade of simple elements of the type of Fig. 1.1, and a cascade of countercurrent units is like a squared-off cascade. The ensuing discussion applies to both square cascades and countercurrent units. Indeed, it is to preserve this generality that the notation of Eq. 2.1 has been introduced.

1. STEADY STATE OF A SQUARE CASCADE (OR COUNTERCURRENT COLUMN)

In the steady state the first integral of Eq. 2.1 is

$$PN_P = PN + c_1N(1 - N) - c_5 \frac{dN}{ds} \tag{2.3}$$

which separates to give a second integral,

$$\int_0^S ds = \int_{N_0}^{N_S} \frac{c_5 \, dN}{-c_1N^2 + (c_1 + P)N - PN_P} \tag{2.4}$$

which is[1]

$$s = \frac{1}{\epsilon \, \Delta(\psi)} \tanh^{-1} \left[\frac{(N_S - N_0) \, \Delta(\psi)}{(N_S + N_0)(1 + \psi) - 2N_S N_0 - 2N_P \psi} \right] \tag{2.5}$$

where

$$\begin{aligned} \psi &\equiv P/c_1 \\ \Delta(\psi) &\equiv [1 + 2\psi(1 - 2N_P) + \psi^2]^{1/2} \\ \epsilon &\equiv c_1/2c_5 \end{aligned} \tag{2.6}$$

The use of ϵ in Eq. 2.5 agrees with Eq. 2.2; for countercurrent columns in which the constants c_1 and c_5 are fundamental, ϵ is understood to be defined by $\epsilon = c_1/2c_5$. The dimensionless quantity ψ is thought of as a normalized rate of production.

Equation 2.5 gives the distribution in the cascade, for given N_P and P, as a function of s. At the top of the cascade, $N_S = N_P$ and $s = S$ = number of stages (or length of column, depending on the definition and units of ϵ). Equation 2.5 becomes

[1]Alternative expressions for s can be found from the identity

$$2 \tanh^{-1} x = \ln \frac{1 + x}{1 - x}$$

$$S = \frac{1}{\epsilon\,\Delta(\psi)} \tanh^{-1}\left[\frac{(N_P - N_0)\,\Delta(\psi)}{(N_P - 2N_P N_0 - N_0) - (N_P - N_0)\psi}\right] \tag{2.7}$$

If values are assigned to N_P and ψ, Eq. 2.7 is an explicit equation which determines the necessary number of stages (or length of a countercurrent column). If S is given, Eq. 2.7 is an implicit relation between N_P and ψ.

When $P = 0$, then $\psi = 0$, $\Delta(\psi) = 1$, and Eq. 2.5 becomes

$$s = \frac{1}{\epsilon} \tanh^{-1} \frac{N_s - N_0}{N_s - 2N_s N_0 + N_0} = \frac{1}{2\epsilon} \ln R_s R_0 \tag{2.8}$$

or

$$R_s = R_0 e^{2\epsilon s} \tag{2.9}$$

Equation 2.9 is in no way peculiar to square cascades, since at $P = 0$ all cascades, regardless of their shape, have the same distribution of concentrations, because Eq. 1.27 with $P = 0$ is

$$\frac{dN}{ds} = 2\epsilon N(1 - N) \tag{2.10}$$

or

$$\frac{dR}{ds} = 2\epsilon R \tag{2.11}$$

whence Eq. 2.9 follows.

However, although the distribution in an ideal cascade with production at the designed rate is again exponential (with half the slope of Eq. 2.9),

$$R_s = R_0 e^{\epsilon s} \tag{2.12}$$

square cascades have the much more complicated distribution of Eq. 2.5.

In the special case when both $N_0 \ll 1$ and $N_P \ll 1$, Eq. 2.7 takes the simpler forms

$$S = \frac{1}{2\epsilon(1 + \psi)} \ln \frac{N_P}{N_0 - \psi(N_P - N_0)} \tag{2.13}$$

$$\frac{N_P}{N_0} = \frac{(1+\psi)e^{2\epsilon S(1+\psi)}}{1+\psi e^{2\epsilon S(1+\psi)}} \tag{2.14}$$

Figure 2.1 illustrates the dependence of fractionation on rate of production according to Eq. 2.14 for various values of $e^{2\epsilon s}$.

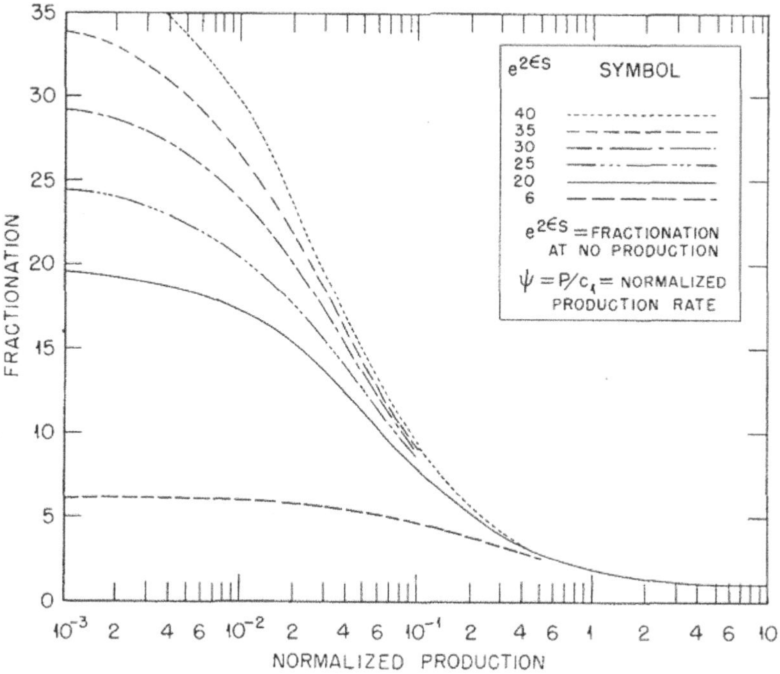

Fig. 2.1—The effect of production on fractionation.

Referring to Eq. 2.3, since the terms $PN + c_1N(1-N)$ increase with N, so also must dN/ds. The smallest value of dN/ds is therefore at the foot of the cascade where N is smallest, $N = N_0$. The maximum transport occurs when this smallest value of $\left.\frac{dN}{ds}\right]_0 = 0$, or

$$\psi = \frac{N_0(1-N_0)}{N_P - N_0} \tag{2.15}$$

For this value of ψ Eq. 2.7 becomes

$$S = \frac{1}{\epsilon\,\Delta(\psi)} \tanh^{-1} 1 = \infty \tag{2.16}$$

Equation 2.7, for a given value of N_P, may be satisfied by an infinite range of pairs of values of S and ψ, ranging from $S = \infty$, $\psi = N_0(1 - N_0)/(N_P - N_0)$, to $S = 1/2\epsilon \ln R_P/R_0$, $\psi = 0$. The optimum length of cascade and rate of production are that pair of values (S,ψ) which makes the average transport per stage P/S a maximum or, what is the same thing, makes $S/c_1\psi$ a minimum. From Eq. 2.7

$$\frac{S}{c_1\psi} = 2\frac{c_5}{c_1^2}\frac{1}{\psi\,\Delta(\psi)}\tanh^{-1}\left[\frac{(N_P - N_0)\,\Delta(\psi)}{(N_P - 2N_PN_0 + N_0) - (N_P - N_0)\psi}\right] \tag{2.17}$$

The right-hand side is now a function of ψ alone. Since differentiating Eq. 2.17 gives a transcendental equation, the minimum is best found by direct computation.

2. SQUARED-OFF CASCADES (CASCADES OF COUNTERCURRENT ELEMENTS)

In the problem of squaring-off an ideal cascade which was previously considered, it was possible to use square sections of any desired length[1] and breadth, i.e., Z and c_5 arbitrary. The usual cascade problem is to make a cascade from elements that are identical countercurrent columns or small-capacity square cascades. Each element has the same constants, c_1 and c_5, and the same length, Z.[1]

Subsequent examples will show that cascades that are very close to ideal may be expected, provided the square elements are sufficiently small.[2]

Instead of solving the problem just set up, i.e., the best N_S with Z_S given, it is much more convenient to proceed indirectly. Suppose that the steps of the cascade, i.e., the N_S, are given, and solve for the best lengths Z_S to fit this cascade. Then it will be seen how the solution of the second problem solves the desired one.

The number of columns in parallel in the *s*th stage is denoted by ν_S. The mole fraction fed into the *s*th stage is N_S, as before. In place of Eq. 2.3, for each of ν columns,

$$\frac{PN_P}{\nu} = \frac{PN}{\nu} + c_1N(1 - N) - c_5\frac{dN}{dZ} \tag{2.18}$$

[1]Use is made of Z instead of S for the length of the square sub-cascades, retaining s as the stage number of the grand cascade.

[2]"Sufficiently small" is here used to mean the transport per element ≪ cascade net transport, and length of stage ≪ ideal cascade length $[c_1N_0(1 - N_0) \ll P(N - N_0)$ and $Z \ll 1/\epsilon \ln R_P/R_0]$.

whose integral is (cf. Eq. 2.5)

$$Z_S = \frac{1}{\epsilon\,\Delta(\phi_S)} \tanh^{-1}\left[\frac{(N_{S+1} - N_S)\,\Delta(\phi_S)}{(N_{S+1} + N_S)\,(1 + \phi_S) - 2N_{S+1}N_S - 2N_P\phi_S}\right] \qquad (2.19)$$

where

$$\phi_S = P/\nu_S\,c_1 \qquad (2.20)$$

$$\Delta(\phi_S) = [1 + 2\phi_S\,(1 - 2N_P) + \phi_S^2]^{1/2}$$

The length of columns in the entire *s*th stage which changes material of mole fraction N_s to material of mole fraction N_{S+1}, while the cascade as a whole produces material of mole fraction N_P at the rate P, is

$$\nu_S Z_S = \frac{2c_5P}{c_1^2}\,\frac{1}{\phi_S\,\Delta(\phi_S)} \tanh^{-1}\left[\frac{(N_{S+1} - N_S)\,\Delta(\phi_S)}{(N_{S+1} + N_S)\,(1 + \phi_S) - 2N_{S+1}N_S - 2N_P\phi_S}\right] \qquad (2.21)$$

$\nu_S Z_S$ is a function of ϕ_S and its minimum (e.g., at $\phi_S = \bar{\phi}_S$) gives the minimum length of column for stage s. From $\bar{\phi}_S$ it is possible to obtain the corresponding optimum value of ν_S,

$$\bar{\nu}_S = \frac{P}{c_1\bar{\phi}_S} \qquad (2.22)$$

the number of columns in parallel, and also $\bar{Z}_S$, the appropriate length of each individual column,

$$\bar{Z}_S = \frac{1}{\epsilon\,\Delta(\bar{\phi}_S)} \tanh^{-1}\left[\frac{(N_{S+1} - N_S)\,\Delta(\bar{\phi}_S)}{(N_{S+1} + N_S)\,(1 + \bar{\phi}_S) - 2N_{S+1}N_S - 2N_P\bar{\phi}_S}\right] \qquad (2.23)$$

3. RAPID APPROXIMATE SOLUTION FOR SQUARED-OFF CASCADES

The solution, Eqs. 2.21 to 2.23, takes a simple form for moderate values of R_{S+1}/R_S. It may be presumed, subject to later check, that the argument of the arctanh is small. Since $\tanh^{-1} x = x + \frac{1}{3}x^3 + \ldots$, neglecting terms of third and higher order,

$$\nu_S Z_S \approx \frac{2c_5P}{c_1^2}\,\frac{N_{S+1} - N_S}{(N_{S+1} - 2N_{S+1}N_S + N_S)\phi_S - (2N_P - N_{S+1} - N_S)\phi_S^2} \qquad (2.24)$$

which is a minimum for

$$\overline{\phi}_S \approx \frac{N_{S+1} + N_S - 2N_{S+1}N_S}{2(2N_P - N_{S+1} - N_S)} \qquad (2.25)$$

With this value of ϕ_S the argument of the arctanh becomes

$$\frac{(N_{S+1} - N_S)\, 2\Delta(\overline{\phi}_S)}{(N_{S+1} - 2N_{S+1}N_S + N_S)} = 2\Delta(\overline{\phi}_S)\frac{R_{S+1} - R_S}{R_{S+1} + R_S}$$

For any stage of a cascade except the top, as $R_{S+1} \to R_S$, $\phi_S \to N_S(1 - N_S)/2(N_P - N_S)$ = finite, and the argument $\to 0$. But for the topmost stage the argument is not small even for R_P very nearly equal to R_S. For, when $N_{S+1} = N_P$,

$$\overline{\phi}_S \approx \frac{R_P + R_S}{2(R_P - R_S)}$$

and, as $R_P \to R_S$, $\Delta(\overline{\phi}_S) \to \overline{\phi}_S$, and the argument $\to 1$. Thus, except for the last stage of a cascade, the approximations in Eqs. 2.24 and 2.25 will be valid for sufficiently small values of R_{S+1}/R_S.

Using the value Eq. 2.25 for $\overline{\phi}_S$, from Eq. 2.24 is obtained

$$\overline{\nu_S Z_S} \approx \frac{8c_5P}{c_1^2}\frac{(2N_P - N_{S+1} - N_S)(N_{S+1} - N_S)}{(N_{S+1} - 2N_{S+1}N_S + N_S)^2} \qquad (2.26)$$

From Eqs. 2.22 and 2.25,

$$\overline{\nu}_S \approx \frac{2P}{c_1}\frac{(2N_P - N_{S+1} - N_S)}{(N_{S+1} - 2N_{S+1}N_S + N_S)} \qquad (2.27)$$

and from Eqs. 2.26 and 2.27,

$$\overline{Z}_S \approx \frac{2}{\epsilon}\frac{N_{S+1} - N_S}{(N_{S+1} - 2N_{S+1}N_S + N_S)} = \frac{2}{\epsilon}\frac{R_{S+1} - R_S}{R_{S+1} + R_S} \qquad (2.28)$$

Solving Eq. 2.28 for R_{S+1}/R_S leads to

$$\frac{R_{S+1}}{R_S} \approx \frac{1 + \frac{\epsilon\overline{Z}_S}{2}}{1 - \frac{\epsilon\overline{Z}_S}{2}} \approx e^{\epsilon\overline{Z}_S} + \frac{(\epsilon\overline{Z}_S)^3}{12} + \ldots \qquad (2.29)$$

A column of length $\overline{Z}_s$ operating with no production will, according to Eq. 2.11, produce an enrichment of $e^{2\epsilon \overline{Z}_s}$. Consequently the theorem has been proved that a column or square cascade in the body of a cascade should be operated at the square root of its maximum fractionation. The foregoing result is really another manifestation of the similar relation found for ideal cascades (Chap. 1, Sec. 2).

Although the approximate expressions (Eqs. 2.25 and 2.26) are fairly inaccurate unless $(R_{s+1} - R_s)/R_s$ is small, experience has shown that the final relation

$$\epsilon \overline{Z}_s = \ln \frac{R_{s+1}}{R_s} \tag{2.30}$$

and the square-root theorem, which follows from it, both hold for $R_{s+1}/R_s \sim 2$.

The recommended rapid procedure for finding optimum lengths and numbers of columns when the N_s are given is therefore as follows: The value of $\overline{Z}_s$ is found from Eq. 2.30. Using this value in Eq. 2.23, $\overline{\phi}_s$ is found. (This gives a better value than Eq. 2.25.) Then $\overline{\nu}_s$ may be found from Eq. 2.22. It should be remarked that even though $\overline{Z}_s$ from Eq. 2.30 will be slightly in error, and likewise $\overline{\nu}_s$, $\overline{\nu_s Z_s}$ will be indistinguishable from the true minimum because of the stationary properties of a minimum.

The solution to the original problem of finding optimum N_s and ν_s when the Z are given follows immediately. The N_s are found from Eq. 2.30, and the rest of the procedure is the same as in the preceding paragraph.

The final proof that the procedure outlined gives a satisfactory answer is the comparison of the resulting cascade with an ideal (infinitesimal-step) cascade. As $N_{s+1} \to N_s$, from Eq. 2.26,

$$\overline{\nu_s Z_s} \to \frac{4c_5 P}{c_1^2} \frac{N_P - N_s}{N_s(1 - N_s)} \frac{N_{s+1} - N_s}{N_s(1 - N_s)} = \frac{4c_5 P}{c_1^2} \frac{N_P - N}{N(1 - N)} \frac{dN}{N(1 - N)} \tag{2.31}$$

and

$$\sum \overline{\nu_s Z_s} \to \frac{4c_5 P}{c_1^2} \int_{N_0}^{N_P} \frac{(N_P - N)dN}{[N(1 - N)]^2} = \frac{4c_5}{c_1^2} PV \tag{2.32}$$

where V is the value function of Eq. 1.43.

Equation 2.32 shows that, for a square cascade or column defined by Eq. 2.31, the separative power per unit length or per stage is

$$\delta U = \frac{c_1^2}{4c_5} \tag{2.33}$$

In the case of the square cascade whose elements are of the type shown in Fig. 1.1, where $c_5 = L/2$ and $c_1 = \epsilon L$, Eq. 2.33 reduces to

$$\delta U = \frac{c_1^2}{4c_5} = \frac{1}{2} L\epsilon^2$$

which coincides with the previous expression for the separative power of such elements, Eq. 1.42.

4. SQUARING-OFF THE TOP OF A CASCADE

In the problem and example of squaring-off treated in the last two sections, a scheme was adopted which would make the squaring-off uniform over the entire cascade. If the cascade is built up of small standard elements, this represents no difficulty. For example, if the whole cascade is assumed to consist of 10,000 centrifuges, then the top stage is made of 9 centrifuges, the twelfth of 28, etc.

Consider a cascade that is ideal up to a mole fraction N_A and then square up to the top. The square section is taken to have the same flow as the ideal section at the point of intersection. It is necessary to compare the properties of an ideal cascade from N_A to N_P with that of a square cascade from N_A to N_P in which the flow is

$$\frac{2P}{\epsilon} \frac{N_P - N_A}{N_A(1 - N_A)}$$

The equilibrium time increases about twenty-five times as rapidly as the total flow and is usually the factor that decides the number of stages that can be squared-off. The squared-off cascade is shorter than the part of the ideal cascade which it replaces.

REFERENCES

Cohen, K., Report A-530, Feb. 4, 1943.

Cohen, K., Packed Fractionating Columns and the Concentration of Isotopes, J. Chem. Phys., 8: 588 (1940).

Cohen, K., Columbia Ser. No. 4L-X28, Dec. 22, 1942 (letter to J. R. Dunning).

Cohen, K., and I. Kaplan, Report A-101, Jan. 28, 1942.

Chapter 3

EQUILIBRIUM TIME OF A SQUARE CASCADE

The equilibrium-time phenomenon seems at first glance to be merely an obstacle to be overcome in successful cascade operation. Indeed, it causes serious production losses in large plants by causing delays when operation is changed, as well as in the initial incubation period. Even for small experimental cascades, it prevents rapid attainment of steady-state conditions and thus slows up the accumulation of data and necessitates long continuous operation for each experiment.

However, it is possible to use the rate of approach to equilibrium as a powerful tool for determining the constants c_1 and c_5 of the fundamental equation (Eq. 2.1), and consequently to turn the disadvantage into an asset. The theory will be expounded below primarily with this end in view. Solutions will be found for those cases which occur experimentally, and the results will be developed with the aid of tables and graphs for immediate application.

The fundamental partial differential equation (Eq. 2.1)

$$c_6 \frac{\partial N}{\partial t} = c_5 \frac{\partial^2 N}{\partial s^2} - \frac{\partial}{\partial s} \left[PN + c_1 N(1 - N)\right]$$

is quasi-linear and parabolic and can be solved by an iteration method. When $N \ll 1$, Eq. 2.1 becomes linear, reducing to

$$c_6 \frac{\partial N}{\partial t} = c_5 \frac{\partial^2 N}{\partial s^2} - (P + c_1) \frac{\partial N}{\partial s} \qquad (3.1)$$

Equation 2.1 also becomes linear even for large mole fractions provided the change in mole fractions is small. The substitution $N = N_0 + x$ gives the linear equation

$$c_6 \frac{\partial x}{\partial t} = c_5 \frac{\partial^2 x}{\partial s^2} - \left[P + c_1(1 - 2N_0)\right] \frac{\partial x}{\partial s}$$

whose solution closely resembles that of Eq. 3.1.

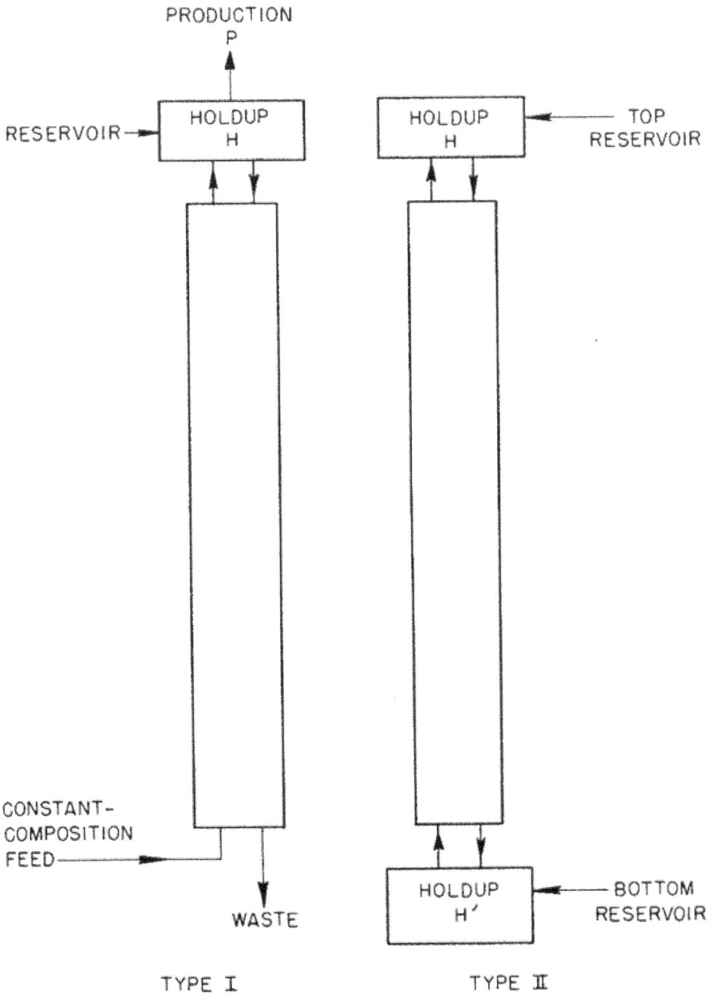

Fig. 3.1—Diagrammatic representation of the two main types of experiment on columns.

Consider first the solution of Eq. 3.1, with $N \ll 1$. (For $N \sim 1$, see Appendix B.)

There are two main types of experiments: In Type I operation (Fig. 3.1) material is withdrawn continuously from one end of the column (by convention the "top"). The other end is maintained at a constant composition by supplying fresh feed continuously, which is equivalent to connecting the bottom of the columns to an infinite reservoir. There is usually a certain amount of external holdup at the product end in connecting piping, surge tanks, etc. The holdup in this top reservoir is denoted H.

In Type II operation (Fig. 3.1) no material is withdrawn. Both ends feed into and out of finite reservoirs. The holdup of the bottom reservoir (at $s = 0$) is called H'. Type I operation with $P = 0$ coincides with Type II operation with $H' = \infty$.

A still more general type of operation (H and H' finite; $P \neq 0$), of which both Types I and II are special cases, is of no physical significance, since the system is gradually emptied of material and there is no steady state.

Before solving Eq. 3.1 it is convenient to divide through by c_5, giving

$$\lambda \frac{\partial N}{\partial t} = \frac{\partial^2 N}{\partial s^2} - 2\epsilon(1 + \psi) \frac{\partial N}{\partial s} \tag{3.2}$$

where

$$\lambda = \frac{c_6}{c_5} \qquad \psi = \frac{P}{c_1} \qquad \epsilon = \frac{c_1}{2c_5} \tag{3.3}$$

The definitions of ψ and ϵ are repeated for convenience. For square cascades of elements of the type illustrated in Fig. 1.1, $\lambda = 2h$ by Eq. 2.2. If s is dimensionless, λ is expressed in units of time. In Type II operation, $\psi = 0$ in Eq. 3.2.

The boundary conditions for Type I operation are[1]

$$\text{At } s = 0, \qquad N = N_0$$

$$\text{At } s = S, \qquad \frac{\partial N}{\partial s} - 2\epsilon N = -K \frac{\partial N}{\partial t} \tag{3.4}$$

$$\text{At } t = 0, \qquad N = N_0$$

[1]The difference in concentration between the reservoirs and the ends of the column may be neglected if $K/\lambda S^2 \ll 1$, which is almost always the case.

The boundary conditions under Type II operation are

$$\text{At } s = 0, \quad \frac{\partial N}{\partial s} - 2\epsilon N = K' \frac{\partial N}{\partial t}$$

$$\text{At } s = S, \quad \frac{\partial N}{\partial s} - 2\epsilon N = -K \frac{\partial N}{\partial t} \tag{3.5}$$

$$\text{At } t = 0, \quad N = N_0$$

In Eqs. 3.4 and 3.5 the factors introduced were

$$K = \frac{H}{c_5}$$
$$K' = \frac{H'}{c_5} \tag{3.6}$$

By far the most expeditious method of solving Eq. 3.2 under these boundary conditions is the use of the Laplace transform. Briefly stated, the method is as follows: By the transformation

$$G(s,p) = \int_0^\infty e^{-pt} N(s,t)\, dt \tag{3.7}$$

the partial differential equation (Eq. 3.2) is transformed into an ordinary differential equation for G with p as a parameter. The solution of the ordinary differential equation gives G(s,p) explicitly. The inverse operation to Eq. 3.7,

$$N(s,t) = \frac{1}{2\pi i} \int_{Br_1} G(s,p) e^{pt}\, dp \tag{3.8}$$

then gives the required solution. In Eq. 3.8, Br_1 is the first Bromwich contour in the complex p plane, a line from $-i\infty$ to $+i\infty$, deformed if necessary to pass to the right of the poles of G(s,p).

Equation 3.2 will now be solved according to the above outline. Multiplying Eq. 3.2 by $e^{-pt}\, dt$ and integrating from 0 to ∞ gives the equation

$$\frac{d^2G}{ds^2} - 2\epsilon(1 + \psi)\frac{dG}{ds} = \lambda(pG - N_0) \tag{3.9}$$

Here again $\psi = 0$ in Type II operation. In establishing Eq. 3.9 the boundary condition at $t = 0$ has been used to evaluate

$$\int_0^{\infty} \frac{\partial N}{\partial t} e^{-pt}\, dt$$

For, integrating by parts,

$$\int_0^{\infty} \frac{\partial N}{\partial t} e^{-pt}\, dt = Ne^{-pt}\Big]_0^{\infty} + p\int_0^{\infty} Ne^{-pt}\, dt = -N_0 + pG$$

The remaining boundary conditions likewise transform. For Type I operation Eq. 3.4 becomes

$$\begin{aligned} &\text{At } s = 0, \qquad G = \frac{N_0}{p} \\ &\text{At } s = S, \qquad \frac{dG}{ds} - 2\epsilon G = -K(pG - N_0) \end{aligned} \tag{3.10}$$

For Type II operation Eq. 3.5 becomes

$$\begin{aligned} &\text{At } s = 0, \qquad \frac{dG}{ds} - 2\epsilon G = K'(pG - N_0) \\ &\text{At } s = S, \qquad \frac{dG}{ds} - 2\epsilon G = -K(pG - N_0) \end{aligned} \tag{3.11}$$

The general solution of Eq. 3.9 is

$$G = \frac{N_0}{p} + C_1 e^{m_1 s} + C_2 e^{m_2 s} \tag{3.12}$$

where C_1 and C_2 are constants and m_1 and m_2 are the solutions of

$$m^2 - 2\epsilon(1 + \psi)m - \lambda p = 0 \tag{3.13}$$

The constants C_1 and C_2 are determined by substituting in Eq. 3.10 for Type I operation or Eq. 3.11 for Type II operation. Omitting these details, the results are as follows.

Type I Operation:

$$G(s,p) = \frac{N_0}{p} + \frac{2\epsilon N_0 e^{\epsilon(1+\psi)(s-S)} \sin \gamma s}{p\gamma \cos \gamma S + p[Kp - \epsilon(1 - \psi)] \sin \gamma S} \tag{3.14}$$

with

$$\gamma \equiv i\beta \equiv i\sqrt{\epsilon^2(1 + \psi)^2 + \lambda p} \tag{3.15}$$

Type II Operation:

$$G(s,p) = \frac{N_0}{p} + \frac{2\epsilon N_0 e^{\epsilon s} \sin \gamma s}{p\gamma} - \frac{f(S,p)\, g(s,p)}{pD(S,p)} \tag{3.16}$$

where the functions f, g, and D are defined by

$$f(S,p) \equiv \frac{2\epsilon N_0}{p}\left[e^{-\epsilon S} + \frac{1}{\gamma}(\epsilon - pK)\sin \gamma S - \cos \gamma S\right]$$

$$g(s,p) \equiv e^{\epsilon s}\left(\cos \gamma s + \frac{pK' + \epsilon}{\gamma} \sin \gamma s\right) \tag{3.17}$$

$$D(S,p) \equiv \frac{(K' - K)\epsilon - pKK' - \lambda}{\gamma} \sin \gamma S - (K + K')\cos \gamma S$$

and γ is the same as in Eq. 3.15, but with $\psi = 0$,

$$\gamma \equiv i\beta \equiv i\sqrt{\epsilon^2 + \lambda p} \tag{3.18}$$

The inversion of either expression for G(s,p) can be carried out in two ways, giving two different infinite series for the same function N(s,t). The difference between the two methods is that one series converges rapidly for small values of t, the other for large values. Although either expansion can be used over the entire range of t by carrying enough terms, it is more convenient to use both expressions, each in its appropriate range. It is possible in this way to find N(s,t) with considerable accuracy for all values of t, using only the first two terms of the series.

1. N(s,t) FOR VALUES OF $t > \lambda S^2/10$

The inverse of G(s,p) suitable for calculations for large t is the usual Heaviside expansion. It is obtained by evaluating Eq. 3.8 as the sum of the residues at the poles of G. Carrying out the indicated operations, the following is obtained.

Type I Operation:

$$N(s,t) = \frac{e^{2\epsilon(1+\psi)s} + \psi e^{2\epsilon(1+\psi)S}}{1 + \psi e^{2\epsilon(1+\psi)S}}$$

$$+ \sum_j \frac{\dfrac{2\epsilon \sin \gamma_j s}{p_j \sin \gamma_j S}\, e^{\epsilon(1-\psi)(s-S)}\, e^{p_j t}}{K + \dfrac{\lambda S}{2} + \dfrac{\lambda}{2\gamma_j^2}\left[Kp_j S - \epsilon(1-\psi)S + 1\right]\left[Kp - \epsilon(1-\psi)\right]} \tag{3.19}$$

where the p_j are the roots of

$$\tan \gamma S = \frac{\gamma}{\epsilon(1-\psi) - Kp} \tag{3.20}$$

Type II Operation:

$$\frac{N(s,t)}{N_0} = e^{2\epsilon s} \frac{K - K' + \lambda S}{K' + Ke^{2\epsilon S} + \frac{\lambda}{2\epsilon}(e^{2\epsilon S} - 1)} - \sum_j \frac{f(S,p_j)\, g(s,p_j) e^{p_j t}}{p_j D'(S,p_j)} \tag{3.21}$$

where the p_j are the roots of $D(S,p) = 0$, or

$$\tan \gamma S = \frac{(K + K')\gamma}{\epsilon(K' - K) - pKK' - \lambda} \tag{3.22}$$

and

$$D'(S,p_j) = \left[\frac{d}{dp} D(S,p)\right]_{p=p_j}$$

$$= \frac{\lambda S \cos \gamma_j S}{2\gamma_j^2}\left[(K' - K)\epsilon - p_j KK' - \lambda\right] \tag{3.23}$$

$$+ \frac{\sin \gamma_j S}{\gamma_j}\left\{- KK' \frac{\lambda\left[(K' - K)\epsilon - p_j KK' - \lambda\right]}{2\gamma_j^2} - \frac{\lambda S(K + K')}{2}\right\}$$

Equations 3.19 and 3.21 describe N as a function of stage number and time for square cascades ($N \ll 1$). A discussion of the roots of Eqs. 3.20 and 3.22 is given in Appendix A. It is shown there that all the roots p_j (there are an infinite number of them) are negative. Thus the terms in the sums are transient, and the constant terms represent $N(s,\infty)$. The second root is shown to be approximately

$$p_2 \approx -\frac{9\pi^2}{4\lambda S^2}$$

Since the coefficient of $e^{p_2 t}$ is much smaller than that of $e^{p_1 t}$, the second transient term is negligible when $p_2 t < -2$; that is, when

$$t > \frac{8\lambda S^2}{9\pi^2} = \frac{\lambda S^2}{10}$$

Consequently, for most purposes only the leading transient term need be considered after $t = \lambda S^2/10$.

Equations 3.19 and 3.21 are valid both for strippers and rectifiers. When the cascade or column is acting as a stripper, c_5, P, and S remain positive, but c_1 is negative. Consequently ϵ and ψ are also negative. The same formal solutions hold. It is quite incorrect to analyze a stripper as if it were a rectifier of the same length, since the equations are not symmetrical about $\epsilon S = 0$.

Particular examples of the solutions given by Eqs. 3.19 and 3.21 may be found in the reports listed in the references for this chapter. The most important special case of the two types of experiment, namely $K' = \infty$, $P = 0$, has been considered in detail. This is the case of a column operated under total reflux (no production), with the concentration at the bottom maintained at N_0. The concentration at the top of the column, or in the reservoir at the top, varies with time according to the relation

$$\frac{N(S,t)}{N_0} = e^{2\epsilon S} + \frac{2\epsilon e^{p_1 t}}{p_1\left[K + \frac{\lambda S}{2} + \frac{\lambda}{2\gamma_1^2}(Kp_1 S - \epsilon S + 1)(Kp_1 - \epsilon)\right]} \tag{3.24}$$

neglecting the higher transients. Now

$$p_1 = -\frac{(\gamma_1 S)^2 + (\epsilon S)^2}{\lambda S^2} \tag{3.25}$$

and $\gamma_1 S$ is the smallest root of

$$\tan \gamma S = \frac{\gamma S}{\epsilon S + \frac{K}{\lambda S}\left[(\gamma S)^2 + (\epsilon S)^2\right]}$$

Therefore $\gamma_1 S$ is a function of ϵS and $K/\lambda S$, and p_1 may be exhibited in the form

$$p_1 = -\frac{B(2\epsilon S, K/\lambda S)}{\lambda S^2} \tag{3.26}$$

Likewise

$$\frac{2\epsilon}{p_1\left[K + \frac{\lambda S}{2} + \frac{\lambda}{2\gamma_1^2}(Kp_1 S - \epsilon S + 1)(Kp_1 - \epsilon)\right]} = -A(2\epsilon S, K/\lambda S)(e^{2\epsilon S} - 1) \tag{3.27}$$

Accordingly Eq. 3.24 takes the form

$$\frac{N(S,t)}{N_0} = e^{2\epsilon S} - (e^{2\epsilon S} - 1)\ A(2\epsilon S, K/\lambda S) \exp - \frac{B(2\epsilon S, K/\lambda S)t}{\lambda S^2} \tag{3.28}$$

Tables of the functions A and B for $K/\lambda S$ between 0.0 and 0.5 and for $0<2\epsilon S<1.20$ are given in Table 3.1 and are plotted in Figs. 3.2 and 3.3. A wider range of these functions for the special case $K = 0$ is given in Table 3.2, where values of $2\epsilon S$ ranging from -5.0 to $+4.0$ are covered.

2. N(s,t) FOR VALUES OF $t < \lambda S^2/10$

Of interest here are the inverses of Eqs. 3.14 and 3.16 in the s,t plane near $t = 0$, that is, the expressions for G(s,p) in the s,p plane for p near $+\infty$. Dropping terms of the order $e^{-2\beta S}$ small compared to unity, Eqs. 3.14 and 3.16 may be written as follows.

Type I Operation:

$$\frac{G(s,p)}{N_0} = \frac{1}{p} + \frac{2\epsilon e^{\epsilon(1+\psi)(s-S)}\left[e^{-\beta(S-s)} - e^{-\beta(S+s)}\right]}{Kp^2 + p\beta + p\epsilon(\psi - 1)} \tag{3.29}$$

Type II Operation:

$$\frac{G(s,p)}{N_0} = \frac{1}{p} - 2\epsilon e^{\epsilon s} \frac{\left\{\frac{\beta}{p}\left[e^{-\beta s} + e^{-\beta(2S-s)}\right] + \left(K - \frac{\epsilon}{p}\right)\left[e^{-\beta s} - e^{-\beta(2S-s)}\right]\right\}}{p^2KK' + p[\lambda + \epsilon(K - K')] + p\beta(K + K')}$$

$$+ 2\epsilon e^{\epsilon(s-S)} \frac{\left\{\frac{\beta}{p}\left[e^{-\beta(S-s)} + e^{-\beta(S+s)}\right] + \left(K' + \frac{\epsilon}{p}\right)\left[e^{-\beta(S-s)} - e^{-\beta(S+s)}\right]\right\}}{p^2KK' + P[\lambda + \epsilon(K - K')] + p\beta(K + K')} \tag{3.30}$$

The term β is defined by Eqs. 3.15 and 3.18. This form is obtained by writing

$$\sin \gamma s = \frac{1}{2}(e^{i\gamma s} - e^{-i\gamma s}) = \frac{1}{2}(e^{-\beta s} - e^{\beta s})$$

and so on.

Equations 3.29 and 3.30 hold to within a few per cent provided $\beta S \approx \sqrt{\lambda p}\, S > \sqrt{10}$; that is, replacing p by 1/t,

$$t < \frac{\lambda S^2}{10} \tag{3.31}$$

Thus they hold in precisely the range of t where the Heaviside expansion begins to become inconvenient.

The inversion of Eqs. 3.29 and 3.30 as they stand, in all their generality, is difficult and not illuminating. Instead, consideration will

Table 3.1—Functions for the Time Equation

$$\frac{N(S,t)}{N_0} = e^{2\epsilon S} - A(e^{2\epsilon S} - 1)e^{-Bt/\lambda S^2}$$

$2\epsilon S$ = ln of steady-state enrichment

$$\frac{K}{\lambda S} = \frac{\text{holdup in reservoir}}{\text{holdup in column}}$$

K/λS	0.0	0.1	0.2	0.3	0.4	0.5
2εS			A(2εS,K/λS)			
0.00	0.810	0.874	0.912	0.934	0.952	0.963
0.10	0.819	0.881	0.918	0.939	0.955	0.966
0.20	0.828	0.887	0.923	0.943	0.958	0.968
0.30	0.836	0.893	0.927	0.947	0.961	0.970
0.40	0.844	0.899	0.931	0.950	0.963	0.971
0.50	0.852	0.904	0.935	0.953	0.965	0.972
0.60	0.859	0.908	0.938	0.956	0.967	0.974
0.80	0.873	0.917	0.945	0.961	0.971	0.978
1.00	0.886	0.926	0.951	0.967	0.976	0.981
1.20	0.899	0.935	0.957	0.972	0.980	0.984
2εS			B(2εS,K/λS)			
0.00	2.468	2.042	1.727	1.488	1.304	1.160
0.10	2.368	1.955	1.649	1.421	1.244	1.105
0.20	2.272	1.870	1.573	1.353	1.185	1.050
0.30	2.181	1.789	1.502	1.291	1.128	1.001
0.40	2.091	1.707	1.432	1.228	1.072	0.952
0.50	2.003	1.631	1.363	1.170	1.021	0.905
0.60	1.921	1.559	1.300	1.112	0.970	0.859
0.80	1.758	1.417	1.178	1.005	0.874	0.773
1.00	1.610	1.284	1.064	0.905	0.786	0.695
1.20	1.469	1.165	0.959	0.813	0.705	0.622

be given various special cases one by one, thereby reducing the complexity of the equations before inversion. For the moment attention should be confined to the mole fractions at the ends of the column (s = 0 and s = S). These are usually the only measurable mole fractions anyway, and the restriction greatly simplifies the inversions.[1]

[1]The rest of the column will be considered in Sec. 3.

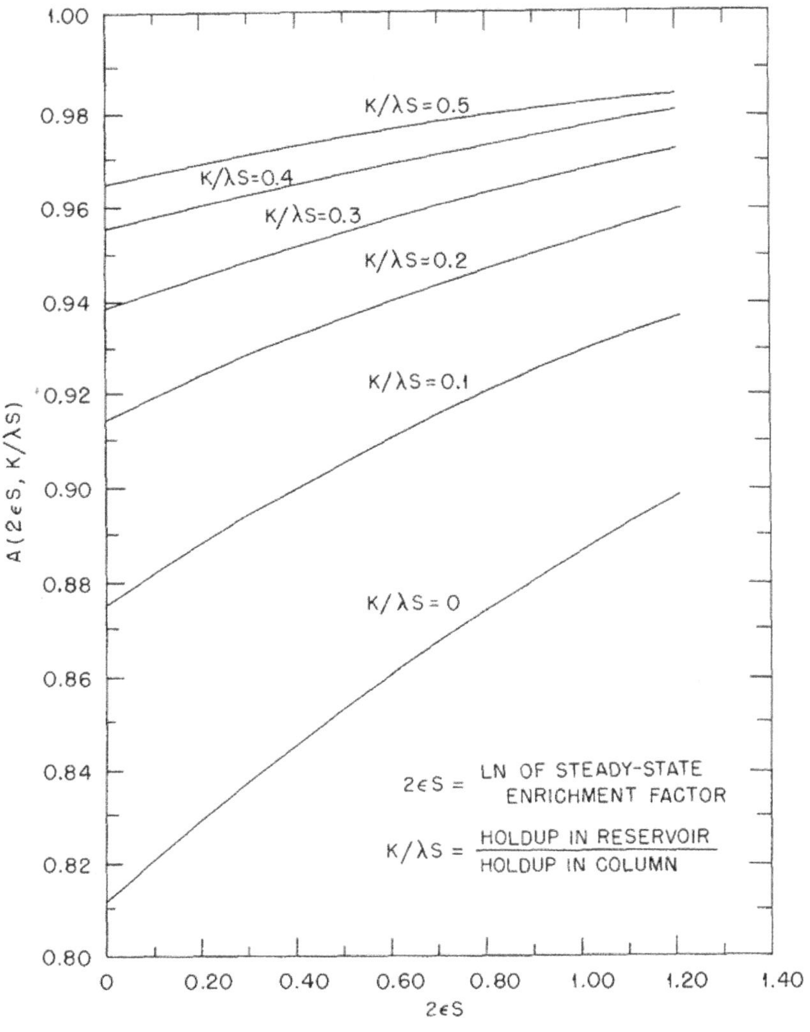

Fig. 3.2—A function for the time equation $A(2\epsilon S, K/\lambda S)$.

First consider special cases of Type II operation.

Type II, Case I: $K = K' = 0$ (no reservoirs at either end)

$$\text{At } s = S, \qquad \frac{G}{N_0} = \frac{1}{p} + \frac{2\epsilon\left(\sqrt{\lambda p + \epsilon^2} + \epsilon\right)}{\lambda p^2}$$

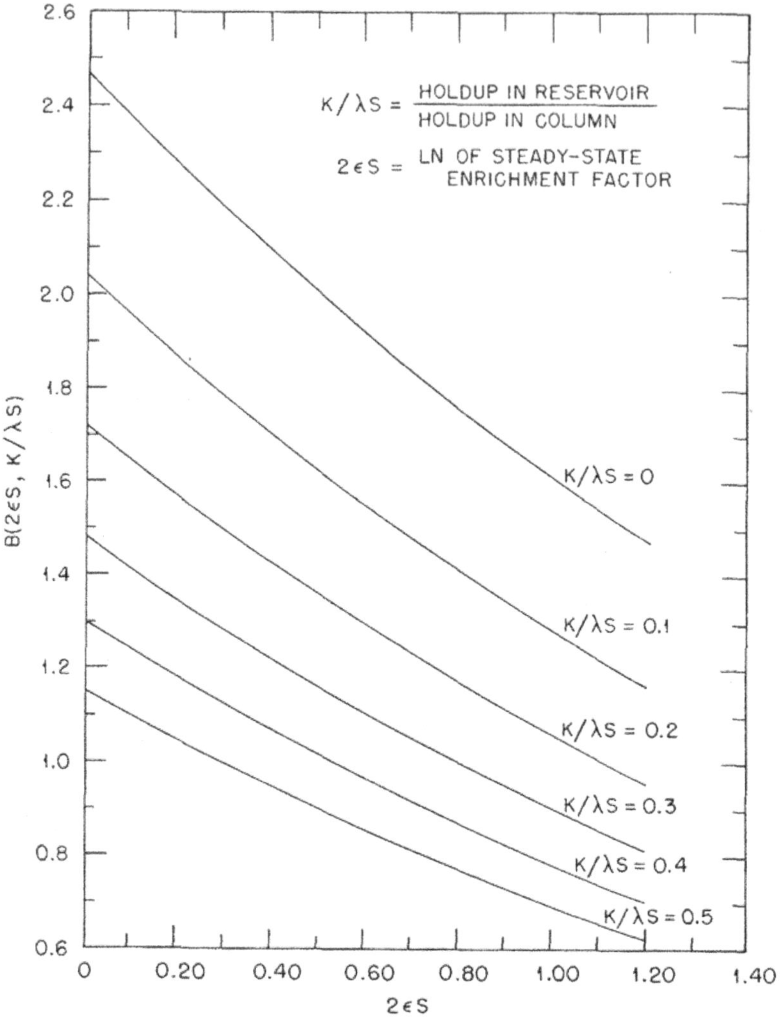

Fig. 3.3—A function for the time equation B(2εS,K/λS).

whence[1]

$$\frac{N(S,t)}{N_0} = 1 + 2\frac{\epsilon^2 t}{\lambda} + 4\epsilon\left(\frac{t}{\lambda\pi}\right)^{1/2} M(-\tfrac{1}{2}, \tfrac{3}{2}; -\epsilon^2 t/\lambda) \tag{3.32}$$

[1]For tables of inverses, see G. A. Campbell and R. M. Foster, Fourier Integrals for Practical Application, Bell Telephone System Monograph B-584 (1931).

Table 3.2 — Functions for Approach to Equilibrium of a Square Cascade with No External Holdup

$$\frac{N(S,t)}{N_0} = e^{2\epsilon S} - A(e^{2\epsilon S} - 1)e^{-Bt/\lambda S^2}$$

Rectifier				Stripper			
$2\epsilon S$	$e^{2\epsilon S}$	$A(e^{2\epsilon S} - 1)$	$\frac{B}{(2\epsilon S)^2}$	$2\epsilon S$	$e^{2\epsilon S}$	$A(e^{2\epsilon S} - 1)$	$\frac{B}{(2\epsilon S)^2}$
0.40	1.49	0.4150	13.067	0	1	0	
0.50	1.65	0.5529	8.023	−0.2	0.8187	−0.1437	66.83
0.60	1.82	0.7055	5.335	−0.4	0.6703	−0.2552	18.08
0.70	2.01	0.87	3.747	−0.6	0.5488	−0.3407	8.668
0.80	2.23	1.071	2.747	−0.8	0.4493	−0.4044	5.263
0.90	2.46	1.28	2.073	−1.0	0.3679	−0.4514	3.625
1.00	2.72	1.522	1.610	−1.2	0.3012	−0.4846	2.705
1.10	3.00	1.78	1.270	−1.4	0.2466	−0.5064	2.133
1.20	3.32	2.085	1.021	−1.6	0.2019	−0.5193	1.750
1.30	3.67	2.41	0.8275	−1.8	0.1653	−0.5257	1.478
1.40	4.06	2.78	0.6818	−2.0	0.1353	−0.5263	1.279
1.50	4.48	3.18	0.5675	−2.2	0.1108	−0.5222	1.128
1.60	4.95	3.64	0.4753	−2.4	0.09072	−0.5144	1.010
1.70	5.47	4.15	0.4010	−2.6	0.07427	−0.5048	0.9150
1.80	6.05	4.71	0.3403	−2.8	0.06081	−0.4930	0.8383
1.90	6.69	5.33	0.2938	−3.0	0.04979	−0.4794	0.7758
2.00	7.39	6.00	0.2500	−3.2	0.04076	−0.4651	0.7228
2.10	8.17	6.75	0.2155	−3.4	0.03337	−0.4500	0.6778
2.20	9.03	7.60	0.1868	−3.6	0.02732	−0.4347	0.6395
2.40	11.02	9.57	0.1418	−3.8	0.02237	−0.4192	0.6065
2.60	13.46	11.98	0.1083	−4.0	0.01832	−0.4036	0.5778
2.80	16.45	14.95	0.08425	−4.2	0.01500	−0.3882	0.5525
3.00	20.09	18.57	0.06550	−4.4	0.01228	−0.3730	0.5298
3.20	24.53	22.99	0.05150	−4.6	0.01005	−0.3583	0.5100
3.40	29.96	28.40	0.04075	−4.8	0.008230	−0.3440	0.4925
3.60	36.60	35.02	0.03250	−5.0	0.006738	−0.3299	0.4768
3.80	44.70	43.10	0.02575				
4.00	54.60	52.98	0.02075				

with $M(-\frac{1}{2}, \frac{3}{2}; -\epsilon^2 t/\lambda)$ a confluent hypergeometric function. Values of the function are listed in Table 3.3.

$$\text{At } s = 0, \qquad \frac{G}{N_0} = \frac{1}{p} - \frac{2\epsilon\left(\sqrt{\lambda p + \epsilon^2} - \epsilon\right)}{p^2}$$

whence

$$\frac{N(0,t)}{N_0} = 1 + 2\,\frac{\epsilon^2 t}{\lambda} - 4\epsilon\left(\frac{t}{\lambda\pi}\right)^{1/2} M\left(-\tfrac{1}{2}, \tfrac{3}{2}; -\epsilon^2 t/\lambda\right) \qquad (3.33)$$

Equations 3.32 and 3.33 are the desired expressions.

Type II, Case II: $K' = \infty$, $K = 0$ (infinite reservoir at bottom, no reservoir on top)

Table 3.3 — The Confluent Hypergeometric Function $M(-\frac{1}{2},\frac{3}{2};-x)$

$$M(-\tfrac{1}{2},\tfrac{3}{2};-x) = \frac{1}{2}e^{-x} + \frac{1}{2}\left(x + \frac{1}{2}\right)\sqrt{\frac{\pi}{x}}\ \text{erf}\ \sqrt{x}$$

x	M
0.0	1.000
0.2	1.065
0.4	1.128
0.6	1.189
0.8	1.247
1.0	1.304
1.2	1.359
1.4	1.412
1.6	1.464
1.8	1.514
2.0	1.563
2.2	1.611
2.4	1.657
2.8	1.746
3.0	1.790
3.5	1.895
4.0	1.995

At $s = S$,
$$\frac{G}{N_0} = \frac{1}{p} + \frac{2\epsilon}{p(-\epsilon + \sqrt{\lambda p + \epsilon^2})} = \frac{1}{p} + \frac{2\epsilon\left(\sqrt{\lambda p + \epsilon^2} + \epsilon\right)}{\lambda p^2}$$

whence

$$\frac{N(S,t)}{N_0} = 1 + \frac{2\epsilon^2 t}{\lambda} + 4\epsilon\left(\frac{t}{\lambda\pi}\right)^{1/2} M(-\tfrac{1}{2},\tfrac{3}{2};-\epsilon^2 t/\lambda) \qquad (3.34)$$

At $s = 0$,
$$\frac{G}{N_0} = \frac{1}{p}$$

whence

$$\frac{N(0,t)}{N_0} = 1 \qquad (3.35)$$

Simple expressions for more complicated cases can only be found by further specialization. Appreciable simplification of all formulas would be obtained if ϵ^2 with respect to λp could be neglected. The condition for validity of this procedure to at least 2.5 per cent is

$$t < \frac{1}{(2\epsilon S)^2}\,\frac{\lambda S^2}{10} \qquad (3.36)$$

The condition in Eq. 3.36 is more stringent than in Eq. 3.31 if $2\epsilon S > 1$; but if $2\epsilon S \leq 1$ it is automatically satisfied when Eq. 3.31 holds.

Type II, Case III: $K' = \infty$, K = finite; $2\epsilon S < 1$ (infinite reservoir at bottom, finite reservoir on top; short column)

$$\text{At } s = S, \qquad \frac{G}{N_0} = \frac{1}{p}\,\frac{2\epsilon}{p^2K + p\sqrt{\lambda p}}$$

whence[1]

$$\frac{N(S,t)}{N_0} = 1 + 4\epsilon\left(\frac{t}{\lambda\pi}\right)^{1/2} + \frac{2\epsilon K}{\lambda}\left(\exp\frac{\lambda t}{K^2}\,\mathrm{erfc}\sqrt{\frac{\lambda t}{K}} - 1\right) \tag{3.37}$$

$$\text{At } s = 0, \qquad \frac{G}{N_0} = \frac{1}{p}$$

whence

$$\frac{N(0,t)}{N_0} = 1 \tag{3.38}$$

Type II, Case IV: K, K' both finite; $2\epsilon S < 1$ (finite reservoirs at both ends; short column)

$$\text{At } s = S, \qquad \frac{G}{N_0} = \frac{1}{p} + \frac{2\epsilon}{Kp^{3/2}(p^{1/2} + \sqrt{\lambda}/K)}$$

whence

$$\frac{N(S,t)}{N_0} = 1 + 4\epsilon\left(\frac{t}{\lambda\pi}\right)^{1/2} + \frac{2\epsilon K}{\lambda}\left(\exp\frac{\lambda t}{K^2}\,\mathrm{erfc}\sqrt{\frac{\lambda t}{K}} - 1\right) \tag{3.39}$$

$$\text{At } s = 0, \qquad \frac{G}{N_0} = \frac{1}{p} - \frac{2\epsilon}{K'p^{3/2}(p^{1/2} + \sqrt{\lambda}/K')}$$

whence

$$\frac{N(0,t)}{N_0} = 1 - 4\epsilon\left(\frac{t}{\lambda\pi}\right)^{1/2} - \frac{2\epsilon K'}{\lambda}\left(\exp\frac{\lambda t}{K'^2}\,\mathrm{erfc}\sqrt{\frac{\lambda t}{K'}} - 1\right) \tag{3.40}$$

[1] $\mathrm{erfc}\, x = \frac{2}{\sqrt{\pi}}\int_x^\infty e^{-y^2}\, dy$

Some cases of Type I operation will now be considered. In the most general case,

$$\text{At } s = S, \qquad \frac{G}{N_0} = \frac{1}{p} + \frac{2\epsilon}{Kp^2 + p\sqrt{\epsilon^2(1+\psi)^2 + \lambda p} + p\epsilon(\psi - 1)} \tag{3.41}$$

$$\text{At } s = 0, \qquad \frac{G}{N_0} = \frac{1}{p}$$

whence

$$\frac{N(0,t)}{N_0} = 1 \tag{3.42}$$

If $2\epsilon(1+\psi)S<1$, so that $\epsilon^2(1+\psi)^2$ with respect to λp may be neglected, the result is precisely the same expressions for G as in Type II, Case III. To show the essential difference between operation with production and operation without, the following case is given:

Type I, Case I: $\lambda = 0$ (no holdup in the column itself)

$$\text{At } s = S, \qquad \frac{G}{N_0} = \frac{1}{p} + \frac{2\epsilon}{Kp^2 + 2\epsilon\psi p}$$

whence

$$\frac{N(S,t)}{N_0} = 1 + \frac{1}{\psi}(1 - e^{-2\epsilon\psi/K}\ t) \tag{3.43}$$

N(0,t) is given by Eq. 3.42.

3. DESCRIPTIVE MATERIAL ON RELAXATION PHENOMENA FOR COLUMNS

If the results of the preceding two sections are combined, the approach to equilibrium of square cascades over the entire period of operation may be followed. Figures 3.4 and 3.5 illustrate Type II, Case II, which is a cascade of arbitrary length, with no reservoir at the top and an infinite reservoir at the bottom, operating under total reflux. Plotted in these graphs is $\frac{N(S,t) - N_0}{N_0}$ against $\left(\frac{t}{\lambda S^2}\right)^{1/2}$ for various values of $2\epsilon S = \ln \frac{N(S,\infty)}{N_0}$. The initial parts of the curves (for $t/\lambda S^2 < 0.1$) are plotted from Eq. 3.34, and the subsequent sections

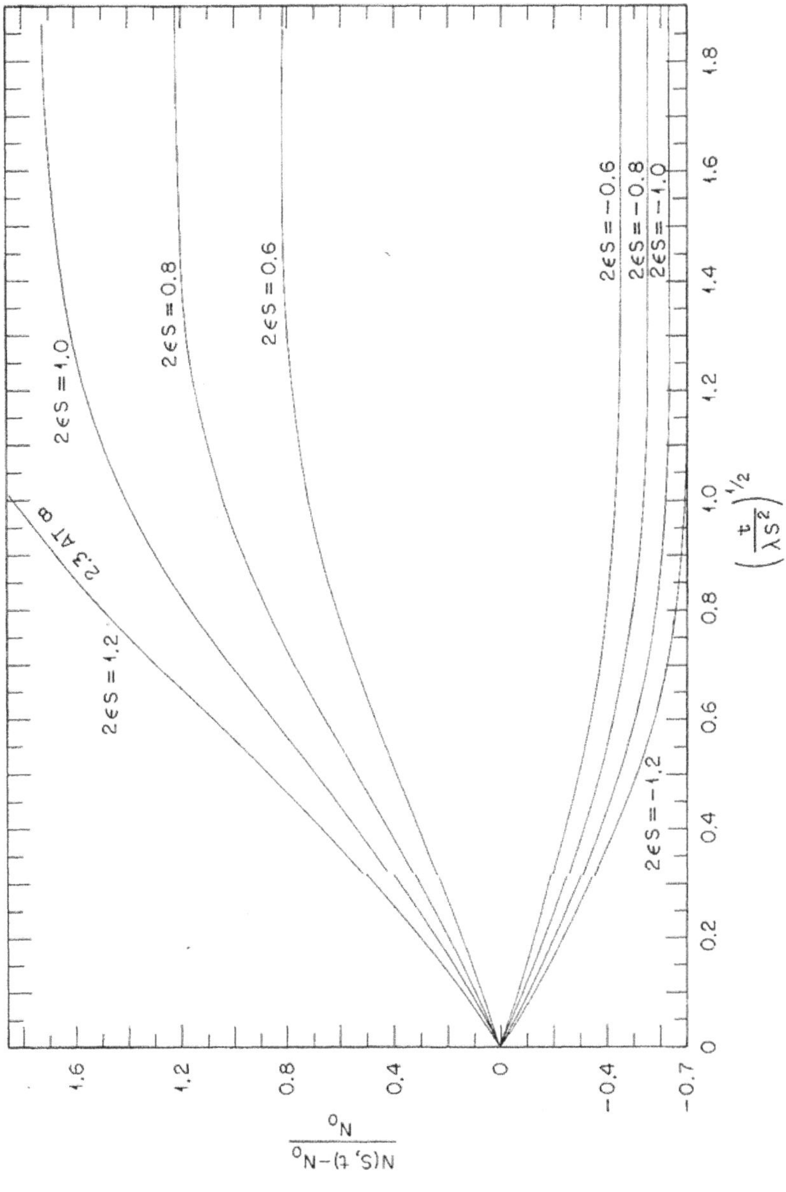

Fig. 3.4—Curves for a cascade of arbitrary length (no reservoir at top; infinite reservoir at bottom). The positive values of $2\epsilon S$ are for rectifier; the negative values are for stripper.

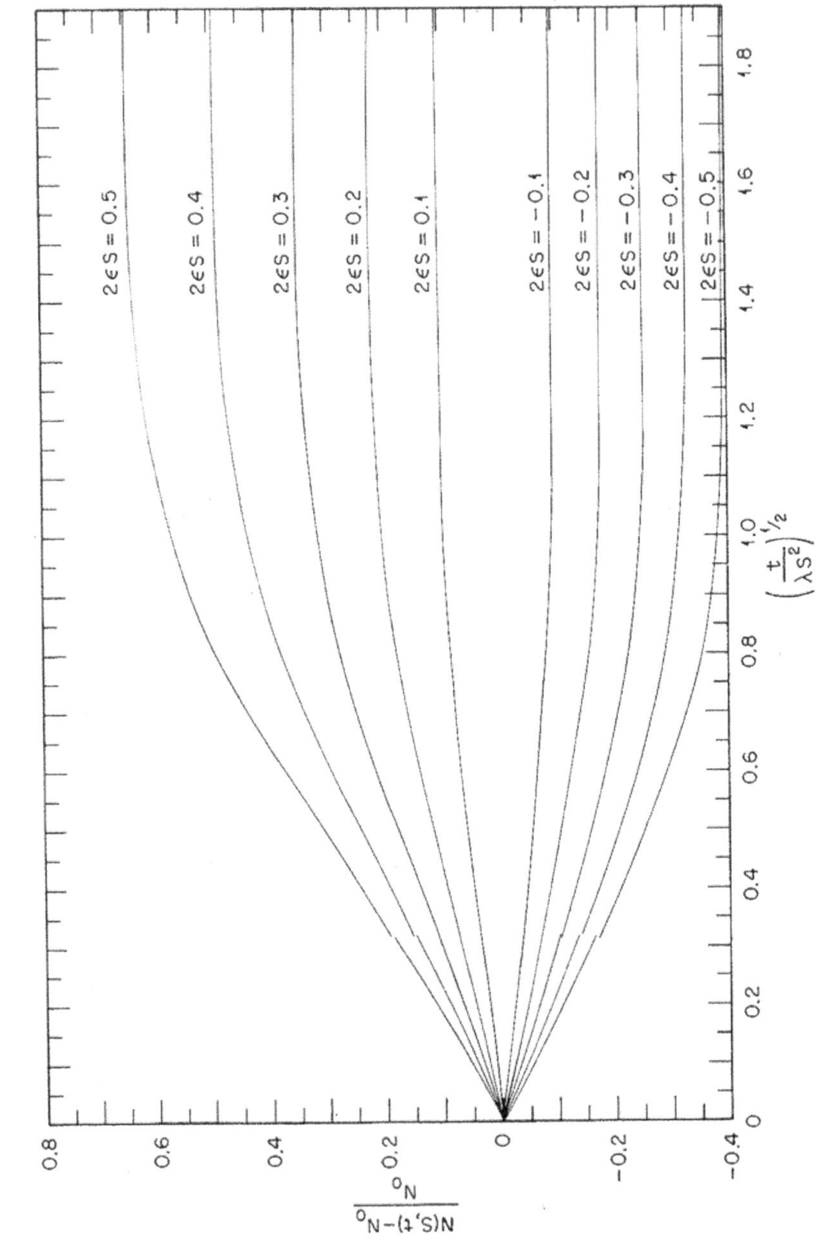

Fig. 3.5—Curves for a cascade of arbitrary length (no reservoir at top; infinite reservoir at bottom). The positive values of $2\epsilon S$ are for rectifier; the negative values are for stripper.

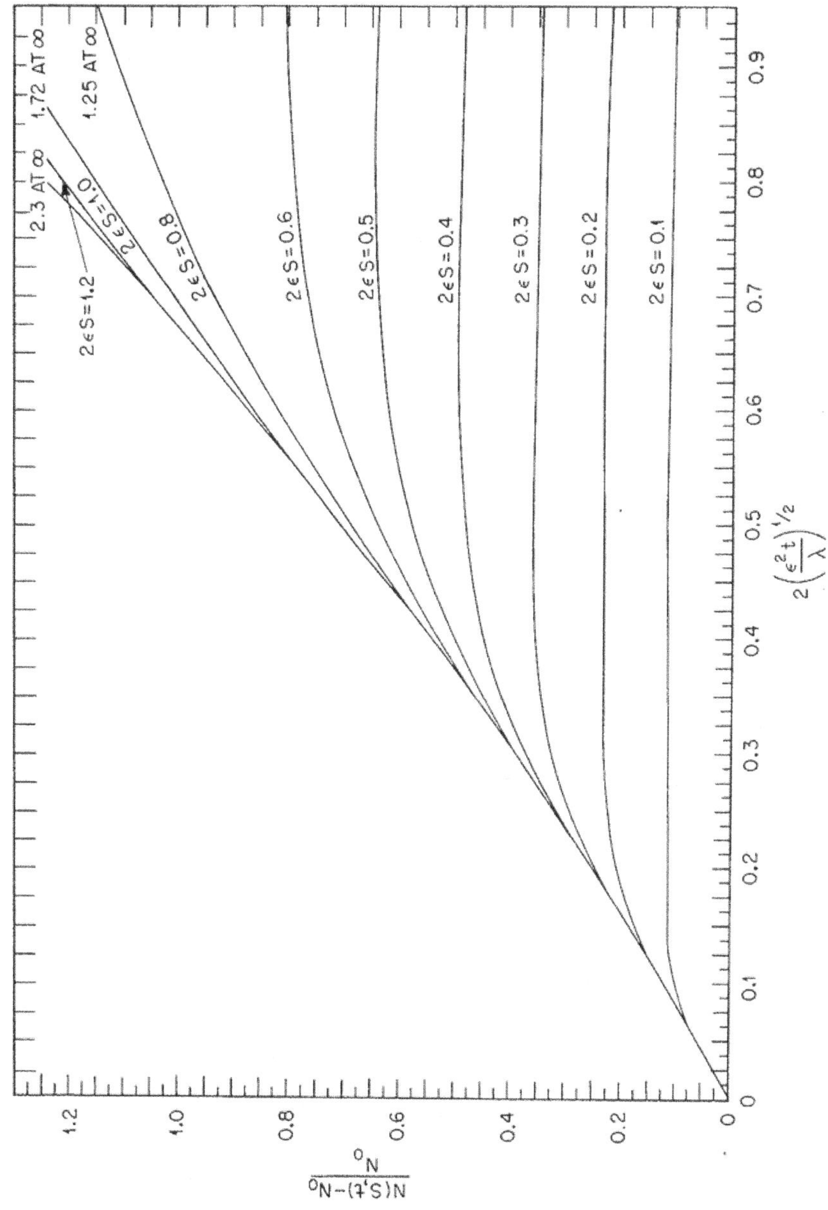

Fig. 3.6—Curves showing similarity of approach toward equilibrium values of $2\epsilon S$. The cascade is of arbitrary length (rectifier) (no reservoir at the top; infinite reservoir at the bottom).

come from the two-term Heaviside expansion (Eq. 3.28). The two different parts of the curve coalesce quite satisfactorily. The curves are straight lines near the origin, the deviation appearing sooner on this scale for long columns (large $2\epsilon S$) than for short columns.

Another way of presenting the same results may be obtained by noting that Eq. 3.34 gives $N(S,t)/N_0$ as a function of $\epsilon^2 t/\lambda$. The expression breaks down at $t/\lambda S^2 \approx 0.1$, or $\epsilon^2 t/\lambda \approx (2\epsilon S)^2/40$. Figure 3.6,

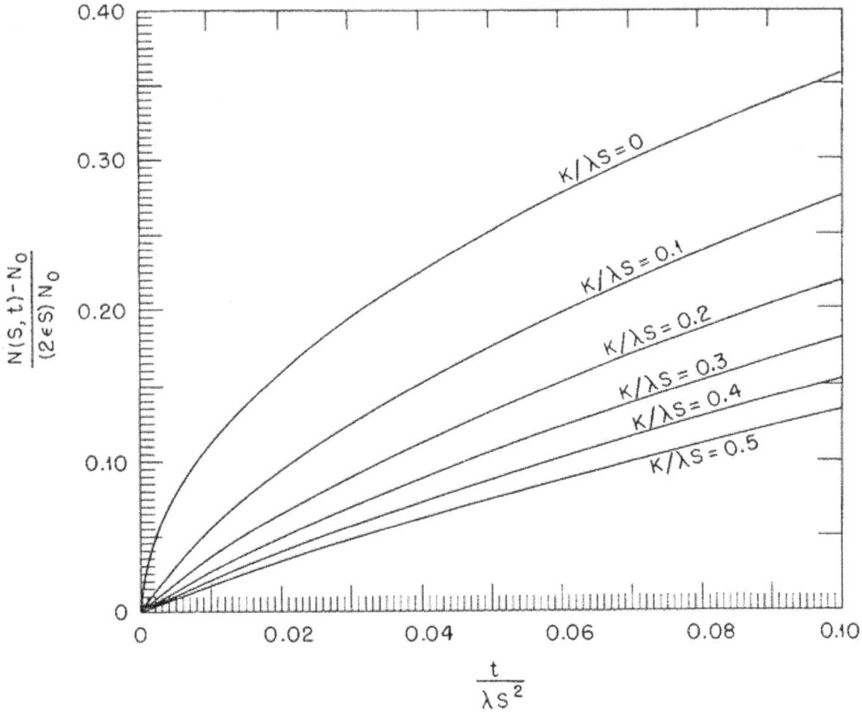

Fig. 3.7—Graph of a short column or cascade (finite reservoir at top; infinite reservoir at bottom).

which gives $\frac{N(S,t) - N_0}{N_0}$ vs. $2\left(\frac{\epsilon^2 t}{\lambda}\right)^{1/2}$, shows how columns with differing $2\epsilon S$ follow the same curve until they break away to approach their equilibrium values.

Figure 3.7 gives a graph of Eq. 3.37 (infinite reservoir at the bottom, finite reservoir at the top) for short columns. This is Type II, Case III, or Type I for $2\epsilon S(1 + \psi) < 1$. Here $\frac{N(S,t) - N_0}{(2\epsilon S)N_0}$ has been plotted against $t/\lambda S^2$ for differing values of the ratio

$$\frac{K}{\lambda S} = \frac{\text{holdup in top reservoir}}{\text{holdup in column}}$$

At high ratios this gives practically a straight-line graph. (But compare with Figs. 3.4 and 3.5, where the abscissa is the square root of $t/\lambda S^2$, and a straight line is found when there is no holdup on top.) Obviously N increases less and less rapidly as K increases.

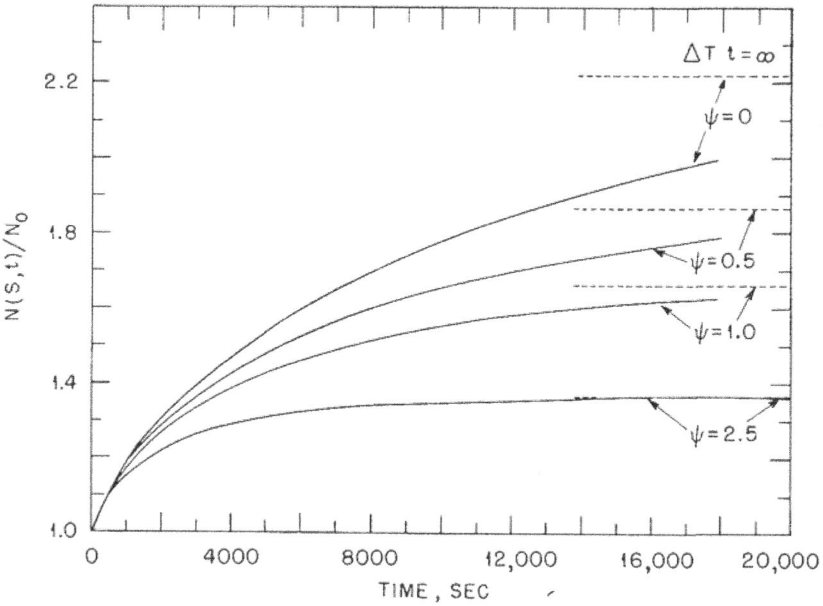

Fig. 3.8—Variation of enrichment with production. Constants: $2\epsilon S = 0.8$, $S = 100$ cm, $\lambda = 2$ sec/sq cm, $K = 0$.

In Fig. 3.8 the effect of changing rate of production on equilibrium time is shown. Here $2\epsilon S = 0.8$, $\lambda = 2$ sec/sq cm, $S = 100$ cm, and $K = 0$. For this graph the solution of Eq. 3.19 was used, keeping two transient terms. Increasing the production rate decreases both the eventual enrichment and the relaxation time of the column. A cascade operating at a lower production rate will reach any particular value of the enrichment sooner than one operating at a higher production rate, except in the very earliest stages of the operation, in which the concentration increase is independent of ψ.

It is of interest to follow theoretically the development of the concentrations along the entire column, especially since it is very diffi-

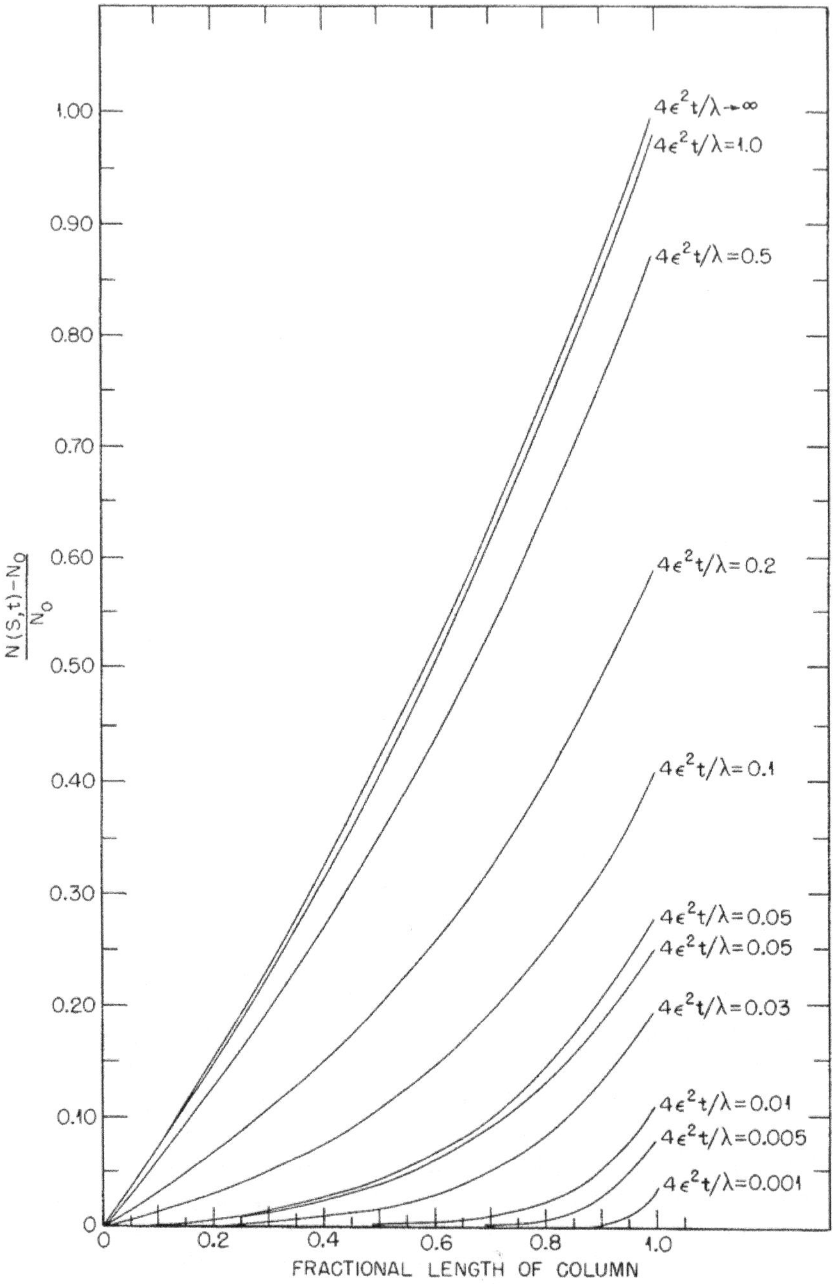

Fig. 3.9—Cascade of fixed length ($2\epsilon S = \ln 2$) (infinite reservoir at the bottom; no reservoir at the top).

cult to do so experimentally. Referring, therefore, to the master equations (Eqs. 3.29 and 3.30), an infinite reservoir is taken at the bottom, no reservoir on top, and $2\epsilon S < 1$ (Type II, Case II for short columns), and it is found that

$$\frac{G(s,p)}{N_0} = \frac{1}{p} + \frac{2\epsilon e^{\epsilon(s-S)}\left[e^{-\sqrt{\lambda p}\,(S-s)} - e^{-\sqrt{\lambda p}\,(S+s)}\right]}{p\sqrt{\lambda p}} \qquad (3.44)$$

The inverse of Eq. 3.44 is

$$\frac{N(s,t)-N_0}{N_0} = e^{-\epsilon S(1-s/S)}\left\{4\sqrt{\frac{\epsilon^2 t}{\pi\lambda}}\exp-\frac{[2\epsilon S(1-s/S)]^2}{16\epsilon^2 t/\lambda} - 2\epsilon S(1-s/S)\right.$$

$$\operatorname{erfc}\frac{2\epsilon S(1-s/S)}{4\sqrt{\epsilon^2 t/\lambda}} - 4\sqrt{\frac{\epsilon^2 t}{\pi\lambda}}\exp-\frac{[2\epsilon S(1+s/S)]^2}{16\epsilon^2 t/\lambda}$$

$$\left. + 2\epsilon S(1+s/S)\operatorname{erfc}\frac{2\epsilon S(1-s/S)}{4\sqrt{\epsilon^2 t/\lambda}}\right\} \qquad (3.45)$$

which holds as long as $\epsilon^2 t/\lambda < (2\epsilon S)^2/40$. For large values of $\epsilon^2 t/\lambda$ Eq. 3.28 may be used with $K = 0$.

Figure 3.9 shows the behavior of such a column whose equilibrium fraction is twofold. Thus $2\epsilon S = \ln 2 = 0.693$, and the critical value of $\epsilon^2 t/\lambda$ is 0.0125. Here $\frac{N(s,t)-N_0}{N_0}$ is plotted as a function of s/S for various values of $4\epsilon^2 t/\lambda$. The upper curves ($\epsilon^2 t/\lambda \geq 0.0125$) were plotted from Eq. 3.28, the lower ($\epsilon^2 t/\lambda \leq 0.0125$) from Eq. 3.45. The two sets of curves fit admirably. It should be added that, during the early stages of the experiment, enriched material accumulates only at the top of the column; when this enriched slug has extended itself down to the foot of the column, the nature of the process changes, and there is a change from one form of asymptotic equation to the other.

REFERENCES

Cohen, K., Report A-530, Feb. 4, 1943.

Cohen, K., and I. Kaplan, Report A-195, June 17, 1942; Report A-395, Nov. 19, 1942; Report A-1276, May 1, 1944; Report A-1292, May 29, 1944.

Kaplan, I., Columbia Ser. No. 4L-X136, Oct. 12, 1942 (letter to C. A. Hutchison).

Chapter 4

DETERMINATION OF CASCADE CONSTANTS

A complete analysis of certain transient phenomena of square cascades was given in the last chapter. This amount of material would not be justified were its purpose purely descriptive. The real need for the detailed results just obtained is given in this chapter, in which it is shown how to turn the equilibrium effects into a tool for determining the cascade constants.

It is always desirable to determine the fundamental constants of the steady state, c_1 and c_5, by direct experiment. In some cases, such as in thermal diffusion, the constants cannot be predicted because of ignorance of the theory. In other cases, such as centrifuges, in which the constants are known functions, subsidiary parameters (viscosity, diffusion coefficients, etc.) are not known to sufficient accuracy. In all cases it is useful to have experimental confirmation that cascade operation follows calculation.

This restriction is mainly responsible for the elegance of the results—the generalization to $N \sim 1$ will be much more cumbersome. Take the holdup per stage or per unit length, c_6, the production, P, and the number of stages, S, as known; c_1 and c_5 are to be determined. Consider first the evaluation of the constants along conventional lines, i.e., by experiments on the steady state.

Equation 2.13, which may be written

$$2\epsilon S = \frac{1}{1 + \dfrac{P}{c_1}} \ln \frac{N_P}{N_0 - \dfrac{P}{c_1}(N_P - N_0)}$$

contains the two parameters ϵ and c_1, which ostensibly may be determined by any two observations of P and N_P. Such a naive procedure is subject to considerable experimental error, and another method, allowing the use of many data, will be presented.

An experiment should be considered in which the withdrawal of product at the top is balanced by the introduction of an equal amount of feed at the same place (Fig. 4.1). In this way the flows up and down are equal, so that the operation of the unit resembles that of a column in the base stages of a cascade; and the equations are appreciably simplified.[1]

Fig. 4.1—An experiment for the determination of column constants.

Following out the same kind of analysis as in Chap. 1, Secs. 2 and 6, this case gives, analogous to Eq. 2.3,

$$2\epsilon\psi(N_P - N_0) = 2\epsilon N - \frac{dN}{ds} \tag{4.1}$$

[1]This device furnished by W. I. Thompson.

The solution of Eq. 4.1 is

$$\frac{N_P}{N_0} - 1 = \frac{e^{2\epsilon S} - 1}{1 + \psi(e^{2\epsilon S} - 1)} \tag{4.2}$$

(compare with Eq. 2.14).

Now, set up for this system the function

$$W = P\left(\frac{N_P}{N_0} - 1\right)^2 = \frac{P(e^{2\epsilon S} - 1)^2}{\left[1 + \frac{P}{c_1}(e^{2\epsilon S} - 1)\right]^2} \tag{4.3}$$

which is a generalization of the work function of Eq. 1.60, defined for infinitesimal concentration changes. The function is plotted against P, from the experimental values of N_P/N_0 and P. The maximum value of W is at

$$\overline{P} = \frac{c_1}{e^{2\epsilon S} - 1} \tag{4.4}$$

and the maximum is

$$\overline{W} = \frac{c_1}{4}(e^{2\epsilon S} - 1) \tag{4.5}$$

It is worth noting that at $P = \overline{P}$

$$\left[\frac{N_P}{N_0} - 1\right]_{P=\overline{P}} = \frac{e^{2\epsilon S} - 1}{2} = \frac{1}{2}\left[\frac{N_P}{N_0} - 1\right]_{P=0} \tag{4.6}$$

The values of $\overline{P}$ and $\overline{W}$ are chosen from the curve. Then

$$c_1 = 2\sqrt{\overline{P}\overline{W}} \tag{4.7}$$

and

$$2\epsilon = \frac{c_1}{c_5} = \frac{1}{S}\ln\left(1 + \frac{c_1}{\overline{P}}\right) \tag{4.8}$$

For small enrichments, $e^{2\epsilon S} \approx 1 + 2\epsilon S$, and the maximum of W is

$$\overline{W} \approx \frac{c_1^2}{4c_5} S \tag{4.9}$$

which gives directly the separative work. At the same time there is experimental determination of the optimum production rate and fractionation.

Next, consideration should be given to the methods based on the equilibrium phenomenon. These are chiefly useful in cases where the time to reach complete equilibrium is of the order of days or weeks, so that a series of steady-state experiments would be excessively time-consuming.

To illustrate the use of relaxation phenomena for determining c_1 and c_5, consider an experiment of Type II, Case III ($K' = \infty$, K = finite, $2\epsilon S < 1$).[1] The pertinent equation for short times is (Eq. 3.37)

$$\frac{N(S,t)}{N_0} = 1 + 4\epsilon\left(\frac{t}{\lambda\pi}\right)^{1/2} \frac{2\epsilon K}{\lambda}\left(\exp\frac{\lambda t}{K^2}\ \operatorname{erfc}\sqrt{\frac{\lambda t}{K}} - 1\right)$$

$$= 1 + \frac{4\epsilon S}{\sqrt{\pi}}\left(\frac{t}{\lambda S^2}\right)^{1/2} + 2\epsilon S\,\frac{K}{\lambda S}\left[\exp\left(\frac{\lambda S}{K}\right)^2 \frac{t}{\lambda S^2}\ \operatorname{erfc}\frac{\lambda S}{K}\left(\frac{t}{\lambda S^2}\right)^{1/2} - 1\right]$$

H, the holdup in the top reservoir, is known, as well as c_6. Thus $\lambda S/K = c_6 S/H$ is known, but ϵ, λ, and K individually are unknown.

Equation 3.37 holds until $t/\lambda S^2 \approx 0.1$. Unless $\lambda S/K \gg 1$ (holdup in column much greater than holdup in reservoir), the argument of the error function

$$\frac{\lambda S}{K}\left(\frac{t}{\lambda S^2}\right)^{1/2}$$

will be small as long as the equation is valid. Expanding the error function in ascending powers of the argument, Eq. 3.37 behaves like

$$\frac{N(S,t)}{N_0} = 1 + \frac{2\epsilon t}{K} \tag{4.10}$$

The deviation from this formula becomes less and less as $t \to 0$.

On the other hand, should the holdup in the reservoir be negligible (e.g., $\lambda S/K > 10$), then unless $t/\lambda S^2$ is very small the argument of the erfc will be large,

$$\operatorname{erfc}\frac{\lambda S}{K}\left(\frac{t}{\lambda S^2}\right)^{1/2} \approx \frac{K}{\lambda S}\left(\frac{\lambda S^2}{\pi t}\right)^{1/2} \exp -\left(\frac{\lambda S}{K}\right)^2 \frac{t}{\lambda S^2}$$

[1]The sample size is supposed to be negligibly small.

and Eq. 3.37 behaves like

$$\frac{N(S,t)}{N_0} = 1 + 4\epsilon\left(\frac{t}{\lambda\pi}\right)^{1/2} - \frac{2\epsilon K}{\lambda} \tag{4.11}$$

over a considerable region.

Depending on which is the appropriate case, $\frac{N(S,t)}{N_0} - 1$ is plotted against t or $t^{1/2}$. The slope of the straight-line portion of the curve is found, which is either $2\epsilon/K = c_1/H$, or $4\epsilon/\sqrt{\lambda\pi} = 2c_1/\sqrt{\pi c_5 c_6}$. In either case a relation for the desired constants is obtained.

The data for long times, which are more abundant, are worked up in quite a different manner. The pertinent equation is (Eq. 3.28)

$$\frac{N(S,t)}{N_0} = e^{2\epsilon S} - A(e^{2\epsilon S} - 1)e^{-Bt/\lambda S^2}$$

Here the procedure is to find $e^{2\epsilon S}$ by extrapolation and then to plot $\ln\left[e^{2\epsilon S} - \frac{N(S,t)}{N_0}\right]$ against t. If the initial choice of $e^{2\epsilon S}$ is good, this graph will be a straight line[1] whose intercept is $\ln[A(e^{2\epsilon S} - 1)]$. (A is found from Fig. 3.2.) If this is not the case, a slightly different choice of $e^{2\epsilon S}$ must be made. From the straight line finally obtained, p_1 is found as the slope.

A consistent set of values of $2\epsilon S$ and p_1 is thus found. From the former, $c_1/c_5 = 2\epsilon$. From p_1 and the relation

$$c_5 = \frac{-p_1 S^2 c_6}{B(2\epsilon S, K/\lambda S)}$$

where B is found from Fig. 3.3, c_5 is determined.

The weak point of this procedure is the choice of $e^{2\epsilon S}$, the equilibrium fractionation, which is not always unambiguous. However, the value of the separative power of the cascade per stage, which is the really important property of the cascade for design purposes,

$$\frac{c_1^2}{4c_5} = \frac{-(2\epsilon S)^2 p_1 c_6}{4B(2\epsilon S, K/\lambda S)}$$

is unaffected by considerable errors in $2\epsilon S$ when determined in this manner.

[1] Except, of course, for small values of t.

As an example of how simple the whole procedure can be made, take the special case $K = 0$. From a graph of $N(S,t)/N_0$ against t, the equilibrium fractionation may be found, along with $t_{1/2}$, the time when

$$\frac{N(S,t)}{N_0} - 1 = \frac{1}{2}\left[\frac{N(S,\infty)}{N_0} - 1\right]$$

The nomograph (Fig. 4.2) then gives $c_6 c_5/c_1^2$ on the right-hand line when a straightedge is laid against $N(S,\infty)/N_0$ on the left-hand line and $t_{1/2}$ on the center.

1. IDEAL CASCADE WITH LOSSES

So far, cascades, both ideal and otherwise, in which no losses of material occur have been considered. In actual cascades, sources of loss are always present. In most chemical processes the effect of small losses is not serious.

It is obvious that, in a given cascade, as the losses increase the production diminishes, so that a cascade which has losses will have to be bigger to produce the same quantity and quality of material. Consider the problem of constructing a no-mixing cascade that will accommodate losses. To solve it a different procedure[1] from that of Chap. 1 should be employed.

It is necessary to introduce the following notation (see Fig. 4.3):

$$\begin{aligned} \mathcal{L}_S &= \theta_S L_S N_S' = \text{flow of desired isotope in enriched fraction} \\ \mathcal{M}_S &= (1-\theta_S) L_S N_S'' = \text{flow of desired isotope in stripped fraction} \\ l_S &= \theta_S L_S (1 - N_S') = \text{flow of other isotope in enriched fraction} \\ m_S &= (1-\theta_S) L_S (1 - N_S'') = \text{flow of other isotope in stripped fraction} \end{aligned} \tag{4.12}$$

From the conservation of matter,

$$\begin{aligned} \mathcal{L}_S + \mathcal{M}_S &= L_S N_S = \text{flow of desired isotope into the separator} \\ l_S + m_S &= L_S(1 - N_S) = \text{flow of other isotope into the separator} \end{aligned} \tag{4.13}$$

In place of Eq. 1.1,

$$\frac{\mathcal{L}_S}{l_S} = \alpha \frac{\mathcal{L}_S + \mathcal{M}_S}{l_S + m_S} \tag{4.14}$$

[1]Originated by R. P. Feynman.

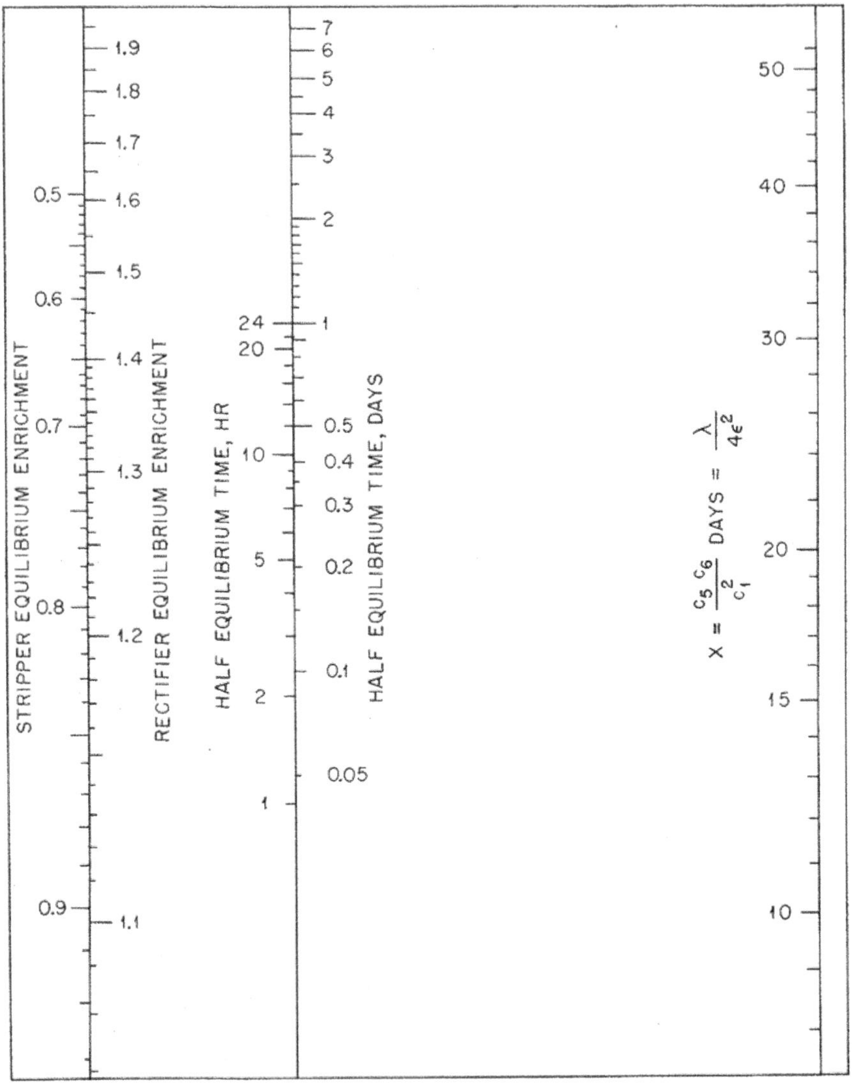

Fig. 4.2—Nomograph for determining column constants.

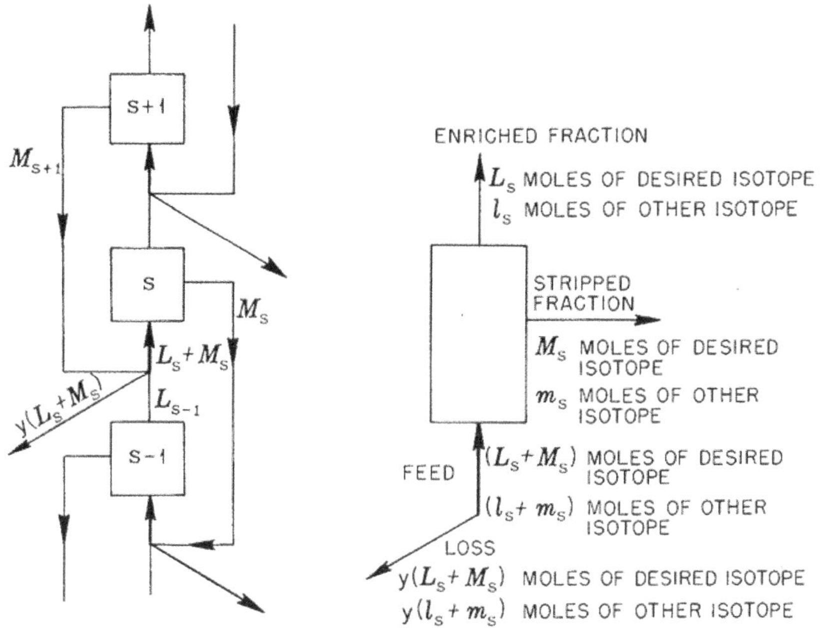

Fig. 4.3—Diagram of a cascade with losses. Only the desired isotope flows are shown. Flows of the other isotope are obtained by using lower-case rather than capital italic letters.

The no-mixing condition leads to the analogues of Eqs. 1.3 to 1.5, the most convenient of which is

$$\frac{L_s}{l_s} = \alpha^2 \frac{M_s}{m_s} \tag{4.15}$$

From Eqs. 4.14 and 4.15, eliminating m_s/l_s,

$$L_s = \alpha M_s \tag{4.16}$$

Now the cascade is connected in such a way (Fig. 4.3) that

$$L_s + M_s = L_{s-1} + M_{s+1} - \text{losses of desired isotope} \tag{4.17}$$

The losses per stage are often proportional to the flow in the stage. Thus the losses from the sth stage of a cascade are set as yL_s, where y is a constant and L_s is as usual the flow entering the stage. Equation 4.17 becomes

$$L_s + M_s = L_{s-1} + M_{s+1} - y(L_s + M_s) \tag{4.18}$$

or, introducing Eq. 4.16,

$$L_{s+1} - (\alpha + 1)(1 + y)L_s + \alpha L_{s-1} = 0 \tag{4.19}$$

Equation 4.19 is a second-order difference equation for L_s. The general solution is

$$L_s = A\omega_1^s + B\omega_2^s \tag{4.20}$$

where ω_1 and ω_2 are the roots of the characteristic equation

$$\omega^2 - (\alpha + 1)(1 + y)\omega + \alpha = 0$$

and can be expressed as

$$\begin{aligned} 2\omega_1(\alpha) &= (\alpha + 1)(1 + y) + \sqrt{(\alpha - 1)^2 + (\alpha + 1)^2(2y + y^2)} \\ 2\omega_2(\alpha) &= (\alpha + 1)(1 + y) - \sqrt{(\alpha - 1)^2 + (\alpha + 1)^2(2y + y^2)} \end{aligned} \tag{4.21}$$

The boundary conditions are

$$\begin{aligned} &\text{At } s = S + 1, \quad L_{S+1} = 0 = A\omega_1^{S+1} + B\omega_2^{S+1} \\ &\text{At } s = S, \quad L_S = PN_P = A\omega_1^S + B\omega_2^S \end{aligned} \tag{4.22}$$

Solving for A and B,

$$L_s = PN_P \left\{ \frac{[\omega_1(\alpha)]^{s-S}}{1 - \dfrac{\omega_1(\alpha)}{\omega_2(\alpha)}} + \frac{[\omega_2(\alpha)]^{s-S}}{1 - \dfrac{\omega_2(\alpha)}{\omega_1(\alpha)}} \right\} \tag{4.23}$$

Now from Eqs. 4.14 and 4.15,

$$l_s = \frac{1}{\alpha} m_s \tag{4.24}$$

while Eq. 4.18 holds with l and m substituted for L and M. Therefore l_s is the same function of $1/\alpha$ that L_s is of α, and in place of Eq. 4.23 the result is

$$l_s = P(1-N_P)\left\{\frac{\left[\omega_1\left(\frac{1}{\alpha}\right)\right]^{s-S}}{1-\frac{\omega_1\left(\frac{1}{\alpha}\right)}{\omega_2\left(\frac{1}{\alpha}\right)}}+\frac{\left[\omega_2\left(\frac{1}{\alpha}\right)\right]^{s-S}}{1-\frac{\omega_2\left(\frac{1}{\alpha}\right)}{\omega_1\left(\frac{1}{\alpha}\right)}}\right\} \tag{4.25}$$

That $\omega_1(1/\alpha) = (1/\alpha)\omega_1(\alpha)$ and $\omega_2(1/\alpha) = (1/\alpha)\omega_2(\alpha)$ is shown by Eq. 4.21. Also since $\omega_1(\alpha)\omega_2(\alpha) = \alpha$, the following schedule may be written:

$$\begin{aligned} \omega_1(\alpha) &= \alpha(1+\Delta) & \omega_1\left(\frac{1}{\alpha}\right) &= 1+\Delta \\ \omega_2(\alpha) &= \frac{1}{1+\Delta} & \omega_2\left(\frac{1}{\alpha}\right) &= \frac{1}{\alpha(1+\Delta)} \end{aligned} \tag{4.26}$$

with Δ defined as

$$\Delta = \frac{1}{2}\left[\frac{\alpha+1}{\alpha}(1+y)+\frac{1}{\alpha}\sqrt{(\alpha-1)^2+(\alpha+1)^2(2y+y^2)}\right]-1 \tag{4.27}$$

Using the form of Eq. 4.27 to simplify L_S and l_S,

$$\begin{aligned} \mathrm{L}_S &= L_S + M_S + l_S + m_S = L_S\left(1+\frac{1}{\alpha}\right)+l_S(1+\alpha) \\ &= \frac{PN_P(\alpha+1)}{\alpha(1+\Delta)^2-1}\Big\{(1+\Delta)^{S-s+2}-\alpha^{s-S-1}(1+\Delta)^{s-S} \\ &\qquad + \frac{1}{R_0}\left[\alpha^{-s}(1+\Delta)^{S-s+2}-\alpha^{-S-1}(1+\Delta)^{s-S}\right]\Big\} \\ &= \frac{2PN_P(\alpha+1)(1+\Delta)}{[\alpha(1+\Delta)^2-1]N_s}\alpha^{(s-S-1)/2}\sinh(S+1-s)\ln(1+\Delta)\sqrt{\alpha} \end{aligned} \tag{4.28}$$

This is the required flow per stage with a fractional loss per stage y.

The total flow in a cascade is obtained by summing the L_S with respect to s from 0 to S. This gives

$$\begin{aligned} \sum_{s=0}^{S}\mathrm{L}_S &= \frac{PN_P(\alpha+1)}{\alpha(1+\Delta)^2-1}\Big\{\frac{(1+\Delta)^{S+1}-1}{\Delta}\left[(1+\Delta)^2-\frac{1}{R_P(1+\Delta)^S}\right] \\ &\qquad + \frac{1-\alpha^{-(S+1)}(1+\Delta)^{-(S+1)}}{\alpha(1+\Delta)-1}\left[\frac{(1+\Delta)^{S+3}\alpha}{R_0}-1\right]\Big\} \end{aligned} \tag{4.29}$$

As before, $R_S = \alpha^S R_0$ and $R_P = \alpha^{S+1} R_0$. When $\Delta \to 0$, Eqs. 4.28 and 4.29 reduce to the familiar expressions of Eqs. 1.11 and 1.13.

Equations 4.28 and 4.29 take more easily understandable forms if $\alpha = 1 + \epsilon$. For then

$$\Delta = -\frac{\epsilon}{2} + \frac{\epsilon}{2}\sqrt{1 + \frac{8y}{\epsilon^2}} \tag{4.30}$$

It is found convenient to define an auxiliary symbol δ by

$$1 + 2\delta = \sqrt{1 + \frac{8y}{\epsilon^2}} \tag{4.31}$$

From Eqs. 4.30 and 4.31, $\delta = \Delta/\epsilon$. Using these relations gives, in place of Eq. 4.28,

$$L_s = \frac{4PN_P}{N_s\,\epsilon(1 + 2\delta)} \exp\left[\frac{\epsilon}{2}(s - S)\right] \sinh\left[\frac{\epsilon}{2}(S - s)(1 + 2\delta)\right] \tag{4.32}$$

and for Eq. 4.29,

$$\sum_{s=0}^{S} L_s = \frac{2PN_P}{\epsilon^2(1 + 2\delta)}\left[\frac{e^{\delta\epsilon S} - 1}{\delta}\left(1 - \frac{1}{R_P e^{\delta\epsilon S}}\right) + \frac{1 - e^{-\epsilon S - \delta\epsilon S}}{1 + \delta}\left(\frac{e^{\delta\epsilon S}}{R_0} - 1\right)\right] \tag{4.33}$$

The relative increase in size of a stage caused by the loss of material is given by the ratio

$$\frac{L_s(\delta)}{L_s(0)} = \frac{1}{1 + 2\delta}\,\frac{\sinh\left[\frac{\epsilon S}{2}(1 + s\delta)\left(1 - \frac{s}{S}\right)\right]}{\sinh\left[\frac{\epsilon S}{2}\left(1 - \frac{s}{S}\right)\right]} \tag{4.34}$$

which is a function of δ. Since δ is in turn a function of $y/\epsilon^2 = yL/L\epsilon^2$, it is seen that the increase in size of a stage is a function of the ratio of loss to separative work. The bottom stages are enlarged proportionally more than the upper stages.

The increase in total flow for a (rectifying) cascade is found by comparing Eq. 4.33 with Eq. 1.23, which is its limit for $y = 0$. Figure 4.4 shows the percentage increase in total flow as a function of y/ϵ^2 for the standard case. It will be seen that a 20 per cent increase in size of the cascade is required when $y = 0.025\,\epsilon^2$.

The above results are quite sensitive to the mole fraction produced by the cascade. Figure 4.5 shows that value of y/ϵ^2 which demands a compensatory increase of 20 per cent in the total flow of the cascade as a function of N_P. A cascade producing a concentration of 10 per cent can afford more than five times the losses of one producing 90

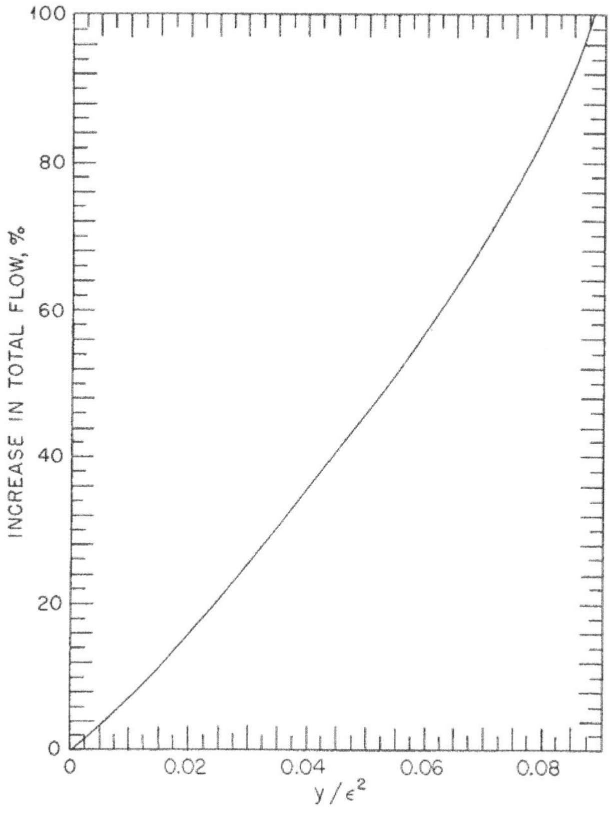

Fig. 4.4—Curve showing the effect of losses on total flow (the loss per stage is proportional to the flow per stage). $N_p = 0.90$, y = fraction of flow lost per stage, ϵ = enrichment per stage.

per cent. Intuitively it should be understood that this effect is due to the increasing penalty for losing high-concentration material on which much work has been done. Obviously it is worth while to take considerable pains to avoid losses in the small upper stages of the cascade. It may well turn out to be economical to use expensive methods and materials in the upper stages if by so doing losses can be cut and thus the amount of overbuilding required in the bulk of the cascade can be diminished.

The effect of losses on a stripping section is somewhat different. The usual objective is not that of producing a given quantity of waste, but rather that of economizing feed. A loss in stage s of the stripper means that a small amount of material is stripped down to N_s and not

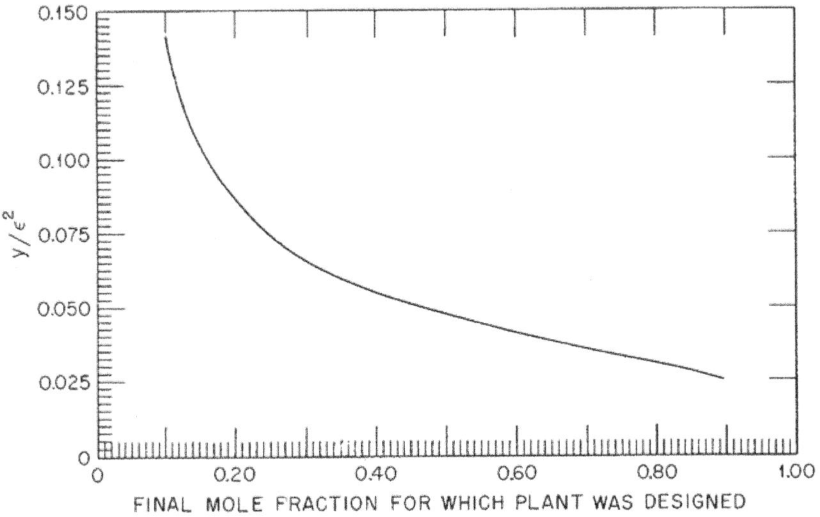

Fig. 4.5—Loss of process material resulting in 20 per cent increase in total flow (the loss per stage is proportional to the flow per stage). The symbol ϵ denotes the enrichment per stage, and y represents the fraction of flow lost per stage.

N_w. The result is to decrease the size of the stripper and increase the required feed. This problem will not be pursued any further here.

The problem solved above (loss proportional to flow) is about the only one that admits a closed analytical solution. But, since in practice the only cases of interest are those in which the effect of losses is small, iteration methods are reasonably successful in dealing with more complicated conditions (see References).

2. COMBINATION OF PLANTS USING DIFFERENT SEPARATION PROCESSES

In this section consideration will be given the connections between two plants with different separation processes, giving expressions with whose aid the factors of plant size and amount of material transferred between the plants may be balanced.

The amount of material transferred between two cascades depends on the mole fraction of the return stream from the upper cascade as follows: Let P_C and N_C be the rate of production and mole fraction of the product from the upper cascade, respectively. Let P_B and N_B be

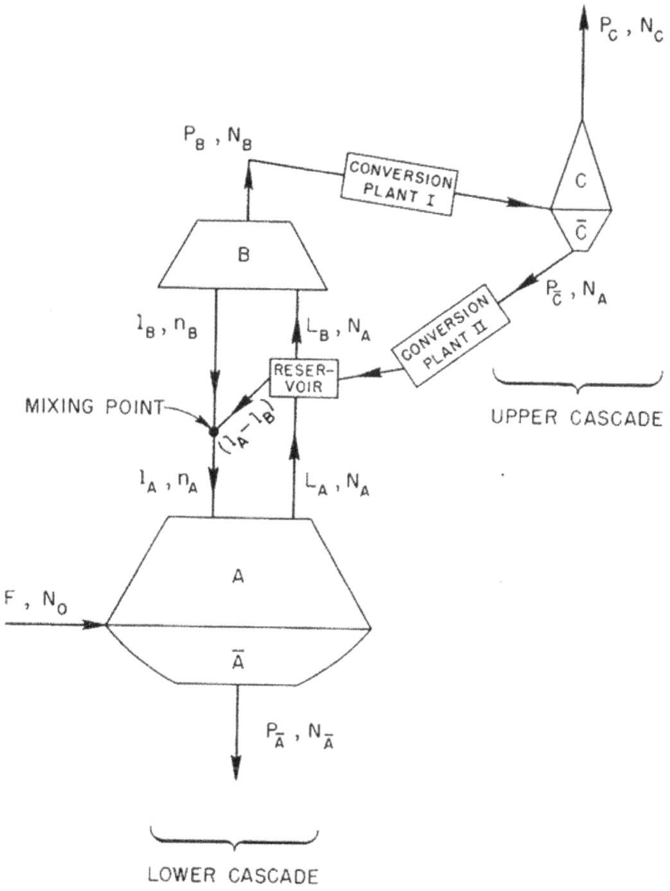

Fig. 4.6—Diagram showing the combination of cascades.

the corresponding quantities for the lower cascade. P_B is at the same time the feed to the upper cascade. $P_{\bar{C}}$ and N_A are the waste stream and its concentration from the upper cascade; $P_{\bar{C}}$ is the return stream to the lower cascade (refer to Fig. 4.6).

The conservation of matter gives the relations

$$P_C = P_B - P_{\bar{C}} \tag{4.35}$$

$$P_C N_C = P_B N_B - P_{\bar{C}} N_A \tag{4.36}$$

From Eqs. 4.35 and 4.36,

$$P_B = P_C \left(1 + \frac{N_C - N_B}{N_B - N_A}\right) \tag{4.37}$$

$$P_{\bar{C}} = P_C \left(\frac{N_C - N_B}{N_B - N_A}\right) \tag{4.38}$$

In these two expressions, P_C and N_C are given (they are the specifications on the final product), and N_B is the mole fraction at which the change from the lower to the upper cascade is made. N_A, the mole fraction of the return stream, is the only variable.

If the upper cascade is an ideal rectifier, Eq. 1.5 gives

$$N_B - N_A = \frac{\alpha - 1}{\alpha} N_B(1 - N_A)$$

For $(\alpha - 1)$ small, $(N_B - N_A)$ is very small, and P_B and $P_{\bar{C}}$ are very large. Thus in general it will be advisable to add a stripper to the upper cascade, as indicated in Fig. 4.6. The connections of the return stream, between parts A and B of the lower cascade, are also shown.

The upper cascade consists of an ordinary rectifier-stripper combination. Cascade B is clearly a rectifier, receiving a feed of material at mole fraction N_A and producing material of mole fraction N_B. From the conservation of matter

$$L_B - l_B = P_B \tag{4.39}$$

$$L_B N_A - l_B n_B = P_B N_B \tag{4.40}$$

The conservation of matter in the reservoir (actually its size is immaterial) requires that

$$L_A + P_{\bar{C}} = L_B + (l_A - l_B) \tag{4.41}$$

or

$$L_A - l_A = (L_B - l_B) - P_{\bar{C}} = P_B - P_{\bar{C}} = P_C \tag{4.42}$$

Furthermore the return stream to cascade A is made up of two parts so that

$$l_A n_A = l_B n_B + (l_A - l_B) N_A$$

Substituting for $l_A - l_B$ from Eq. 4.41,

$$l_A n_A = l_B n_B + (L_A - L_B) N_A + P_{\bar{C}} N_A$$

and rearranging,

$$L_A N_A - l_A n_A = L_B N_A - l_B n_B - P_{\bar{C}} N_A = P_B N_B - P_{\bar{C}} N_A$$

$$L_A N_A - l_A n_A = P_C N_C \tag{4.43}$$

Equations 4.42 and 4.43 show that cascade A is just the part up to N_A of a cascade that produces material (P_C, N_C).

Expressions for the flows and holdups in these sections may easily be found on the assumption that they are ideal cascades. The total flow in cascade section A is

$$\frac{2P_C}{\epsilon_A^2} F_1(N_C, N_A, N_0)$$

if $1 + \epsilon_A$ is the simple-process factor in the lower cascade. The total flow in cascade section B is

$$\frac{2P_B}{\epsilon_A^2} F_1(N_B, N_B, N_A)$$

The total flow in cascade section C is

$$\frac{2P_C}{\epsilon_C^2} F_1(N_C, N_C, N_B)$$

where $1 + \epsilon_C$ is the simple-process factor in the upper cascade.

The total flow in the stripper section $\bar{A}$ is

$$\frac{2P_{\bar{A}}}{\epsilon_A^2} F_1(N_{\bar{A}}, N_{\bar{A}}, N_0)$$

and in the stripper section $\bar{C}$ is

$$\frac{2P_{\bar{C}}}{\epsilon_{\bar{C}}^2} F_1(N_A, N_A, N_B)$$

Analogous expressions hold for the finite case. Expressions for the desired material holdup and the contribution to the equilibrium time of the various sections above the feed point can also be obtained without any trouble (see Appendix C).

Cascades $\bar{A}$, A, and C are the same as the corresponding sections of a single ideal cascade producing (P_C, N_C). But it will be observed that cascade B is not the same as the section between N_A and N_B of a single cascade. B and $\bar{C}$ together are equivalent to (and slightly larger than) this section. An inefficiency in the combined cascades caused by the mixing of the constituents of the return stream to cascade A causes a loss in value of

$$(l_A - l_B)V(N_A) + l_B V(n_B) - l_A V(n_A)$$

which is made up for by the increased size of B and $\bar{C}$.

It will be remarked that a combined cascade, for maximum efficiency, is considerably different from two cascades designed for other duties and connected together. There is a considerable premium on foresight in this problem.

The separation into two cascades which is illustrated here can clearly be carried on further, giving any number of sub-cascades. The chief feature which makes continued subdivision attractive, even when the sub-cascades use identical equipment, is that the flow between cascades (each of which is a rectifier-stripper) is much less than the flow between stages of a single cascade (cf. Eqs. 4.37 and 4.38). This is an advantage in removing accumulated impurities (caused by inleaks, decomposition, etc.) or in control.

REFERENCES

Cohen, K., Report A-60, October 1941; Report A-530, Feb. 4, 1943; Report A-781, Aug. 13, 1943.

Cohen, K., Columbia Ser. No. 4R-X143, May 13, 1942.

Cohen, K., and W. I. Thompson, Columbia Ser. No. 4R-104, Sept. 4, 1943.

Kaplan, I., Columbia Ser. No. 4M-138, Jan. 18, 1944 (memorandum to K. Cohen); Columbia Ser. No. 4M-160, Apr. 17, 1944 (memorandum to M. Kilpatrick).

Lefkowitz, H., and B. Schwartz, Columbia Ser. No. 4R-18, Feb. 4, 1943.

Lefkowitz, H., B. Schwartz, and H. Mayer, Columbia Ser. No. 4R-42, Apr. 12, 1943.

Mayer, H., I. Kaplan, and K. Cohen, Report A-396, Dec. 2, 1942.

Thompson, W. I., Experimental Operation of a Single Centrifuge, S.O.D. Report, June 14, 1943.

Chapter 5

THE CONTROL PROBLEM

The problem of maintaining certain specified variables at preassigned values is common to any planned operation. In a chemical reaction, for example, these quantities are usually the temperature, the pressure, and the amount of material in process. The control problem for isotope-separation plants differs from that of other manufacturing operations in the degree of complexity. The cascades considered in this volume contain thousands of different interdependent stages or elements. Methods of analysis which are successful with systems containing, for example, a dozen related units will fail here. For this reason the control problem for isotope-separation cascades must be treated as a distinct subject.

Furthermore it is often possible to break up a cascade into smaller units. For example, a cascade of 10,000 centrifuges with nine centrifuges in the top stage (Chap. 2, Sec. 4) might be broken up into nine completely independent cascades in parallel, or a cascade might be broken down into sub-cascades along the lines discussed in Chap. 4, Sec. 2. The desirability of such expedients will not be clear until the difficulties of controlling a single large cascade are known. In the discussion that follows, therefore, consideration will be given to a single large rectifier-stripper.

It was shown in Chap. 1, Sec. 2, that for an ideal plant deviations up to 20 per cent from the calculated flows are without appreciable effect on the separation. This also holds true to a considerable extent for nonideal plants of intelligent design. Thus to have a cascade operate according to plan, it is only necessary to see that the flows throughout the plant are approximately correct. This question, of maintaining the temperatures, pressures, flows, etc., near their design values, of seeing that material does not accumulate in or drain out of any part of the plant, and of seeing that small distrubances do not build up to large ones, is one part — the hydrodynamic part — of the control problem.

There is in addition a second part which deals not with the average conditions but with the effects of fluctuations from the average. A slight change in the operating conditions may produce a variety of effects: a change in the transport of material from stage to stage, a mixing of material of different concentrations from different stages, or a change in the separating power of the element. In general, a fluctuation causes mixing, and mixing means a lowering of the concentration of the product.

Both parts of the control problem are essential. Without the hydrodynamic part, the types, magnitudes, and frequencies of variations that are to be expected are unknown. Without a knowledge of the effects of fluctuations on concentration an evaluation cannot be made of the effectiveness of the hydrodynamic control measures so far as plant productivity is concerned.

By its nature, the hydrodynamic problem is intimately associated with the particular separation method, equipment, and control instruments used. Since this problem cannot be generalized without destroying its illustrative value, an explanatory example will be presented.

The problem of the effect of fluctuations on separation is more susceptible of general treatment. Even so there are aspects which differ from one case to another. A cascade of centrifuges, for instance, has certain problems in common with all squared-off cascades; it also has effects due to variations in the behavior of parallel units which a diffusion cascade lacks. But the theory developed in the succeeding sections is basic to all the problems. In Sec. 1 consideration is given to equations governing the change of mole fraction with fluctuations in the operating constants. Their solution in the general case is developed in such a way that the results of the hydrodynamic problem may be used in their most compact form (in the s,p plane of the Laplace transform). In Sec. 2 the average change in transport for periodic fluctuations is derived. Section 3 treats a particular example: the effect of variations in the rate of withdrawal of product on a square cascade. Sinusoidal variations are considered, and the effect of batch (instead of continuous) withdrawal of product is treated.

1. EFFECT OF FLUCTUATIONS ON CONCENTRATION

The fundamental equations of isotope separation as developed in Chap. 1, Sec. 6, using the generalized notations of Chap. 2, are[1]

[1] The terms c_1, c_5, and c_6 were introduced in Chap. 2, Square Cascades, by the schedule in Eq. 2.2, but there is no reason why they cannot be employed as functions of s and t, as indicated.

$$c_1(s,t)\, N(s,t)\, [1 - N(s,t)] - c_5(s,t)\, \frac{\partial N(s,t)}{\partial s} = \tau(s,t) - N(s,t)\, T(s,t) \qquad (5.1)$$

$$\frac{\partial T(s,t)}{\partial s} = -\frac{\partial c_6(s,t)}{\partial t}$$
$$\frac{\partial \tau(s,t)}{\partial s} = -\frac{\partial}{\partial t}[c_6(s,t)\, N(s,t)] \qquad (5.2)$$

The effect of perturbations in the fundamental quantities c_1, c_5, c_6, and P on the separation performed by the cascade is found by solving Eqs. 5.1 and 5.2 with these four quantities preassigned functions of time. Of primary interest are the small fluctuations, which if imposed on a system already in the steady state will give small changes in concentration. Gross upsets and phenomena associated with the approach to equilibrium are excluded from consideration. The desired solution may thus be obtained by ordinary perturbation theory, with the steady state as the unperturbed system.

The quantities c_1, c_5, and c_6 are given as functions of stage number. In cascades with more than one element per stage, they are to be understood as averages over the elements. The mixing losses caused by deviations of parallel elements from their average are not included here.

Suppose that there is a primary disturbance of some nature from an outside source. It will be written as

$$\sigma f(s,t) \qquad (5.3)$$

The parameter σ is not necessarily small—its function is to distinguish between the various orders of terms resulting from the primary disturbance.

As a consequence of σf there will be induced disturbances in c_1, c_5, c_6, and P. Thus c_1, which before the disturbance is a function of s only, becomes a function of s, t, and σ. Expanding in powers of σ,

$$c_1(s,t;\sigma) = c_{10}(s) + \sigma c_{11}(s,t) + \sigma^2 c_{12}(s,t) + \ldots \qquad (5.4)$$

where $c_{10}(s)$ is the undisturbed value of c_1. Generally the symbol $c_1(s,t;\sigma)$ will be shortened to $c_1(s,t)$. Likewise

$$c_5(s,t) = c_{50}(s) + \sigma c_{51}(s,t) + \sigma^2 c_{52}(s,t) + \ldots$$
$$c_6(s,t) = c_{60}(s) + \sigma c_{61}(s,t) + \sigma^2 c_{62}(s,t) + \ldots \qquad (5.5)$$
$$P(t) = P_0 + \sigma P_1(t) + \sigma^2 P_2(t) + \ldots$$

The notation is self-explanatory and defines the coefficients c_{51}, c_{52}, etc. As a result of Eq. 5.5 and the first line of Eq. 5.2,

$$T(s,t) = P_0 + \sigma T_1(s,t) + \sigma^2 T_2(s,t) + \dots \qquad (5.6)$$

The functions $c_1(s,t) \dots T(s,t)$ are considered known, as well as the coefficients of their expansions. They are in fact the solutions of the hydrodynamic part of the control problem.

Now, inserting Eqs. 5.4, 5.5, and 5.6 in Eqs. 5.1 and 5.2, it is desirable to find N(s,t) and $\tau(s,t)$. These will also be functions of σ, which will be written as[1]

$$\begin{aligned} N(s,t) &= N_0(s) + \sigma N_1(s,t) + \sigma^2 N_2(s,t) + \dots \\ \tau(s,t) &= \tau_0 + \sigma \tau_1(s,t) + \sigma^2 \tau_2(s,t) + \dots \end{aligned} \qquad (5.7)$$

Of course

$$\tau_0 = P_0 N_{P_0} \qquad (5.8)$$

where N_{P_0} is the undisturbed mole fraction of the product.

The convergence of Eq. 5.7 is now of concern. The natural period of a cascade with respect to concentration disturbances is the equilibrium time, which we have seen in Chap. 3, Sec. 1, is of the order $\lambda/\epsilon^2 \sim c_{60}c_{50}/c_{10}^2$. Physically this means that Eq. 5.7 will converge provided the average values of the disturbances on most of the stages, over periods of time smaller than λ/ϵ^2, are small compared to the undisturbed quantities.

Introducing the expansions of Eqs. 5.4 to 5.7 into Eqs. 5.1 and 5.2, and separating the coefficients of $\sigma^0, \sigma^1, \sigma^2, \dots,$ the following systems of equations are obtained:

$$P_0 N_{P_0} = P_0 N_0(s) - c_{50}\frac{\partial N_0}{\partial s} + c_{10}N_0(1 - N_0) \qquad (5.9)$$

$$\begin{aligned} \tau_1 &= P_0 N_1 - c_{50}\frac{\partial N_1}{\partial s} + c_{10}(1 - 2N_0)N_1 + T_1 N_0 \\ &\qquad - c_{51}\frac{\partial N_0}{\partial s} + c_{11}N_0(1 - N_0) \\ \frac{\partial \tau_1}{\partial s} &= -c_{60}\frac{\partial N_1}{\partial t} + N_0\frac{\partial T_1}{\partial s} \end{aligned} \qquad (5.10)$$

[1] The symbols $N_0(s)$, $N_1(s,t) \dots$ introduced in Eq. 5.7 are not in contradiction with the previously used symbols of N_s, which have been supplanted by N(s,t) and have no meaning in this continuous treatment. N_F will be used for the feed concentration.

$$\tau_2 = P_0N_2 - c_{50}\frac{\partial N_2}{\partial s} + c_{10}(1 - 2N_0)N_2 + T_1N_1 - c_{51}\frac{\partial N_1}{\partial s}$$

$$+ c_{11}N_1(1 - 2N_0) - c_{10}N_1^2 + T_2N_0 - c_{52}\frac{\partial N_0}{\partial s} + c_{12}N_0(1 - N_0)$$

$$\frac{\partial \tau_2}{\partial s} = -c_{60}\frac{\partial N_2}{\partial t} + N_0\frac{\partial T_2}{\partial s} - c_{61}\frac{\partial N_1}{\partial t} + N_1\frac{\partial T_1}{\partial s} \qquad (5.11)$$

and so on for higher orders.

The boundary conditions for each order equation can likewise be found by expansion. This will be done later in the section (Eqs. 5.23 et seq.). When the boundary conditions are known the equations may be solved successively. In the paragraphs immediately following an investigation will be made of the properties of the differential equations, and method of solving them will be considered.

Equation 5.9 gives the steady state of the unperturbed system. The expression for τ_1 has the form

$$\tau_1 = DN_1 + \Omega_1(s,t) \qquad (5.12)$$

where D is the linear operator, independent of t,

$$D(s) \equiv P_0 - c_{50}\frac{\partial}{\partial s} + c_{10}(1 - 2N_0) \qquad (5.13)$$

and

$$\Omega_1(s,t) = T_1N_0 - c_{51}\frac{\partial N_0}{\partial s} + c_{11}N_0(1 - N_0) \qquad (5.14)$$

is, after substituting $N_0(s)$, which is a solution to Eq. 5.9, an explicit function of s and t. Combining Eq. 5.12 with the second equation of Eq. 5.10 gives

$$c_{60}\frac{\partial N_1}{\partial t} + \frac{\partial}{\partial s}DN_1 = -\frac{\partial \Omega_1}{\partial s} + N_0\frac{\partial T_1}{\partial s} \qquad (5.15)$$

The left-hand side of Eq. 5.15 is a linear homogeneous expression in N_1, with coefficients independent of t; the right-hand side is an explicit function of s and t. It is therefore easily solvable by standard methods. In particular the Laplace transform of Eq. 5.15,

$$c_{60}pN_1^* + \frac{\partial}{\partial s}DN_1^* = -\frac{\partial \Omega_1^*}{\partial s} + N_0\overline{\frac{\partial T_1^*}{\partial s}} \qquad (5.16)$$

is an ordinary linear second-order differential equation in s. It will be remembered that the asterisk (*) was introduced to denote the Laplace transform of any function. Thus if x(s,t) is a function of s and t,

$$x^* = x^*(s,p) = \int_0^\infty e^{-pt}\, x(s,t)\, dt$$

The chief advantage[1] of using the Laplace transform of Eq. 5.15—besides the fact that it solves the equation without trouble in spite of complicated boundary conditions—is that the transforms T_1^*, c_5^*, and c_{11}^* are used on the right-hand side. Since the hydrodynamic part of the control problem will also be solved by using the Laplace transform, these quantities are immediately available. In fact, since the inversion of T_1^* to T_1, etc., is the last and hardest step of the hydrodynamic solution, very considerable difficulties are avoided by using Eq. 5.16.

The solution of Eq. 5.15 or 5.16 may be facilitated by transforming the dependent and independent variables.[2] The expression DN_1 becomes, using the integrating factor,

$$J(s) = \exp\left[-\int_0^s \frac{c_{10}(1 - 2N_0) + P_0}{c_{50}}\, ds\right] \tag{5.17}$$

$$DN_1 = -c_{50} J^{-1} \frac{\partial}{\partial s}(JN_1) \tag{5.18}$$

The dependent variable is then chosen as JN_1. In many cascades the quantities c_{10}, c_{50}, etc., are most conveniently expressed in terms of the N_0. For example, an ideal cascade is defined in terms of the N_0 so that

$$\frac{c_{10}}{c_{50}} = 2\epsilon = \text{constant}$$

$$c_{10} = \frac{2P_0(N_{P_0} - N_0)}{N_0(1 - N_0)}$$

This suggests changing the independent variable from s to N_0, which can be done with the aid of Eq. 5.9.

The expression for τ_2 has the form

[1] This feature was pointed out by H. Mayer.
[2] Following the practice of K. Fuchs.

$$\tau_2 = DN_2 + \Omega_2(s,t) \tag{5.19}$$

where D is the linear operator of Eq. 5.13 and

$$\Omega_2(s,t) = T_1N_1 - c_{51}\frac{\partial N_1}{\partial s} + c_{11}N_1(1-2N_0) - c_{10}N_1^2 + T_2N_0 - c_{52}\frac{\partial N_0}{\partial s} + c_{12}N_0(1-N_0) \tag{5.20}$$

Combining Eq. 5.19 with the second part of Eq. 5.11,

$$c_{60}\frac{\partial N_2}{\partial t} + \frac{\partial}{\partial s}DN_2 = -\frac{\partial \Omega_2}{\partial s} + N_0\frac{\partial T_2}{\partial s} - c_{61}\frac{\partial N_1}{\partial t} + N_1\frac{\partial T_1}{\partial s} \tag{5.21}$$

The Laplace transform of Eq. 5.21 is[1]

$$c_{60}pN_2^* + \frac{\partial}{\partial s}DN_2^* = -\frac{\partial \Omega_2^*}{\partial s} + N_0\frac{\partial T_2^*}{\partial s} - c_{61}\frac{\partial N_1^*}{\partial t} + N_1\frac{\partial T_1^*}{\partial s} \tag{5.22}$$

The process can, at least in principle, be continued indefinitely for N_3, N_4 . . . .

The boundary conditions for the differential equations will now be obtained. First of all the initial condition $N(s,0) = N_0(s)$ leads to

$$N_1(s,0) = N_2(s,0) = \ldots = 0 \tag{5.23}$$

Assume that there are reservoirs at each end of the plant, of holdup H (product end) and H′ (waste end). The product and waste are withdrawn from the reservoirs. The mole fractions in the reservoirs are therefore called $N_P(t)$ and $N_W(t)$, respectively. [Here it is not legitimate to neglect the difference between N_P and $N(S,t)$. A rapid fluctuation in $N(S,t)$, for example, leaves N_P relatively unchanged because of the buffer effect of the holdup.] The waste stream is called $W(t)$. The flow from the cascade into the top reservoir is called $c_5(P,t)$; the flow from the cascade into the bottom reservoir is $c_5(W,t)$. These quantities also have expansions as powers of σ.

[1] The formula for the transform of the product of two functions is

$$[x_1(t)x_2(t)]^* = \frac{1}{2\pi i}\int_{Br_1} x_1^*(p-q)x_2^*(q)\,dq$$

The contour is to the right of the poles of $x_2^*(q)$ and to the left of the poles of $x_1^*(p-q)$.

$$
\begin{aligned}
W(t) &= W_0 + \sigma W_1(t) + \sigma^2 W_2(t) + \dots \\
H &= H_0 + \sigma H_1(t) + \sigma^2 H_2(t) + \dots \\
H' &= H'_0 + \sigma H'_1(t) + \sigma^2 H'_2(t) + \dots \\
N_P(t) &= N_{P_0} + \sigma N_{P_1}(t) + \sigma^2 N_{P_2}(t) + \dots \\
N_w(t) &= N_{w_0} + \sigma N_{w_1}(t) + \sigma^2 N_{w_2}(t) + \dots \\
c_5(P,t) &= c_{50}(P) + \sigma c_{51}(P,t) + \sigma^2 c_{52}(P,t) + \dots \\
c_5(W,t) &= c_{50}(W) + \sigma c_{51}(W,t) + \sigma^2 c_{52}(W,t) + \dots
\end{aligned}
\tag{5.24}
$$

The mole fractions, $N(s,t)$, and the transports, $T(s,t)$ and $\tau(s,t)$, will refer to the stripper if s has negative values. Equations 5.9 to 5.22 hold for the stripper if P_0 and N_{P_0} are replaced throughout by $-W_0$ and N_{W_0}. The stage number of the lowest stage of the stripper will be called "B" (a negative number).

The discontinuity in τ and T at $s = 0$ implied in the change in the equations is accounted for by the introduction of feed (F, N_F).

$$
\begin{aligned}
T(0+,t) - T(0-,t) &= F \\
\tau(0+,t) - \tau(0-,t) &= FN_F \\
N(0+,t) &= N(0-,t)
\end{aligned}
\tag{5.25}
$$

Note, however, that N_F does not have to be the same as $N(0,t)$, or even N_{F_0} the same as $N_0(0)$. F and N_F may both vary with time (the cascade under consideration might be the upper cascade of a combination of plants). Expanding Eq. 5.25 in powers of σ,

$$
\begin{aligned}
P_0 + W_0 &= F_0 \\
P_0 N_{P_0} + W_0 N_{W_0} &= F_0 N_{F_0} \\
N_0(0+) &= N_0(0-)
\end{aligned}
\tag{5.26}
$$

$$
\begin{aligned}
T_1(0+,t) - T_1(0-,t) &= F_1 \\
\tau_1(0+,t) - \tau_1(0-,t) &= F_0 N_{F_1} + F_1 N_{F_0} \\
N_1(0+,t) &= N_1(0-,t)
\end{aligned}
\tag{5.27}
$$

$$
\begin{aligned}
T_2(0+,t) - T_2(0-,t) &= F_2 \\
\tau_2(0+,t) - \tau_2(0-,t) &= F_0 N_{F_2} + F_1 N_{F_1} + F_2 N_{F_0} \\
N_2(0+,t) &= N_2(0-,t)
\end{aligned}
\tag{5.28}
$$

The boundary conditions at the top of the rectifier are obtained from the relations

$$T(S,t) = P(t) + \frac{\partial H}{\partial t}$$

$$\tau(S,t) = P(t)N_P(t) + \frac{\partial}{\partial t}(HN_P) \qquad (5.29)$$

$$c_5(P,t)\,[N(S,t) - N_P(t)] = H\,\frac{\partial N_P}{\partial t}$$

On expanding in powers of σ and collecting, it is found that

$$T_0(S) = P_0$$

$$\tau_0(S) = P_0N_{P_0} \qquad (5.30)$$

$$N_0(S) = N_{P_0}$$

These obvious relations have already been incorporated in the preceding equations (cf. Eqs. 5.6 and 5.8). Likewise, combining the first two equations of Eq. 5.29,

$$\tau_1(S,t) - P_0N_{P_1} = H_0\,\frac{\partial N_{P_1}}{\partial t} + T_1(S,t)N_{P_0}$$

$$c_{50}(P)\,[N_1(S,t) - N_{P_1}(t)] = H_0\,\frac{\partial N_{P_1}}{\partial t} \qquad (5.31)$$

and in the second order,

$$\tau_2(S,t) - P_0N_{P_2} - T_1(S,t)N_{P_1} - T_2(S,t)N_{P_0} = H_0\,\frac{\partial N_{P_2}}{\partial t} + H_1\,\frac{\partial N_{P_1}}{\partial t}$$

$$c_{50}(P)\,[N_2(S,t) - N_{P_2}(t)] + c_{51}(P,t)\,[N_1(S,t) - N_{P_1}(t)] = H_0\,\frac{\partial N_{P_2}}{\partial t} + H_1\,\frac{\partial N_{P_1}}{\partial t} \qquad (5.32)$$

The boundary conditions at the bottom of the stripper are very similar. Corresponding to Eq. 5.29,

$$T(B,t) = -W(t) - \frac{\partial H'}{\partial t}$$

$$\tau(B,t) = -W(t)N_w(t) - \frac{\partial}{\partial t}(H'N_w) \qquad (5.33)$$

$$c_5(W,t)\,[N(B,t) - N_w(t)] = H'\,\frac{\partial N_w}{\partial t}$$

In zero order

$$\begin{aligned} T_0(B) &= -W_0 \\ \tau_0(B) &= -W_0N_{w_0} \\ N_0(B) &= N_{w_0} \end{aligned} \tag{5.34}$$

The first-order terms are

$$\begin{aligned} \tau_1(B,t) + N_{w_1}W_0 &= -H_0' \frac{\partial N_{w_1}}{\partial t} + T_1(B,t)N_{w_0} \\ c_{50}(W)\,[N_1(B,t) - N_{w_1}(t)] &= H_0' \frac{\partial N_{w_1}}{\partial t} \end{aligned} \tag{5.35}$$

and the second-order terms are

$$\begin{aligned} &\tau_2(B,t) + N_{w_2}W_0 - T_1(B,t)N_{w_1} - T_2(B,t)N_{w_0} = H_0' \frac{\partial N_{w_2}}{\partial t} - H_1' \frac{\partial N_{w_1}}{\partial t} \\ &c_{50}(W)\,[N_2(B,t) - N_{w_2}(t)] + c_{51}(W,t)\,[N_1(B,t) - N_{w_1}(t)] \\ &\qquad = H_0' \frac{\partial N_{w_2}}{\partial t} + H_1' \frac{\partial N_{w_1}}{\partial t} \end{aligned} \tag{5.36}$$

Substituting for τ_1 from Eq. 5.12 in the boundary conditions (Eqs. 5.27, 5.31, and 5.35), a complete set of boundary conditions may be found for the first-order differential equation (Eq. 5.25). First an expression for

$$\Delta \frac{\partial N_0(0)}{\partial s} = \frac{\partial N_0}{\partial s}\bigg]_{0+} - \frac{\partial N}{\partial s}\bigg]_{0-}$$

is needed. This can be obtained from Eq. 5.26 by substituting Eq. 5.9 and its analogue with $-W_0$ for P_0 and N_{w_0} for N_{P_0}. It is found that[1]

$$\Delta \frac{\partial N_0(0)}{\partial s} = -\frac{F_0}{c_{50}(0)}\,[N_{F_0} - N_0(0)] \tag{5.37}$$

Then from Eq. 5.27, for the boundary condition at $s = 0$,

[1] $N_{F_0} - N_0(0)$ is found by substituting the solution of Eq. 5.9, giving N_{P_0} in terms of $N_0(0)$, and N_{w_0} in terms of $N_0(0)$, into Eq. 5.26

$$P_0N_{P_0} + W_0N_{w_0} = F_0N_{F_0}$$

$$F_0\,[N_1(0,t) - N_{F_1}(t)] - c_{50}(0)\;\Delta\,\frac{\partial N_1(0,t)}{\partial s} = \left[F_1(t) - \frac{c_{51}(0,t)F_0}{c_{50}(0)}\right](N_{F_0} - N_0) \quad (5.38)$$

From Eq. 5.31,

$$D(S)\,N_1(S,t) - P_0N_{P_1} - H_0\,\frac{\partial N_{P_1}}{\partial t} = c_{51}(S)\,\frac{\partial N_0}{\partial s} - c_{11}(S)N_0(1 - N_0)$$
$$c_{50}(P)\,[N_1(S,t) - N_{P_1}(t)] = H_0\,\frac{\partial N_{P_1}}{\partial t} \quad (5.39)$$

and from Eq. 5.35,

$$D(B)\,N_1(B,t) + W_0N_{W_1} + H_0'\,\frac{\partial N_{W_1}}{\partial t} = c_{51}(B)\,\frac{\partial N_0}{\partial s} - c_{11}(B)N_0(1 - N_0)$$
$$c_{50}(W)\,[N_1(B,t) - N_{W_1}(t)] = H_0'\,\frac{\partial N_{W_1}}{\partial t} \quad (5.40)$$

This elaborate set of conditions is very expeditiously condensed by using the Laplace transform. With the aid of Eq. 5.23,

$$\left(\frac{\partial N_{P_1}}{\partial t}\right)^* = pN_{P_1}^* \qquad \left(\frac{\partial N_{W_1}}{\partial t}\right)^* = pN_{W_1}^*$$

and therefore $N_{P_1}^*$ and $N_{W_1}^*$ may be eliminated altogether by the second parts of Eqs. 5.39 and 5.40, which give

$$N_{P_1}^* = \frac{N_1^*(S,p)}{1 + K_0p}$$
$$N_{W_1}^* = \frac{N_1^*(B,p)}{1 + K_0'p} \quad (5.41)$$

with the definitions

$$K_0 \equiv \frac{H_0}{c_{50}(P)}$$
$$K_0' \equiv \frac{H_0'}{c_{50}(W)} \quad (5.42)$$

as in Chap. 3. With Eq. 5.41, the transforms of Eqs. 5.39 and 5.40 are

$$D(S)\, N_1^*(S,p) - \frac{P_0 + H_0 p}{1 + K_0 p}\, N_1^*(S,p) = c_{51}^*(S)\, \frac{\partial N_0}{\partial s} - c_{11}^*(S) N_0 (1 - N_0) \quad (5.43)$$

$$D(B)\, N_1^*(B,p) + \frac{W_0 + H_0' p}{1 + K_0' p}\, N_1^*(B,p) = c_{51}^*(B)\, \frac{\partial N_0}{\partial s} - c_{11}^*(B) N_0 (1 - N_0) \quad (5.44)$$

The transform of Eq. 5.38 is

$$F_0[N_1^*(0,p) - N_{F_1}^*(p)] - c_{50}(0)\, \Delta \frac{\partial}{\partial s}\, N_1^*(0,p) = \left[F_1^*(p) - \frac{c_{51}^*(0,p)\, F_0}{c_{50}(0)}\right](N_{F_0} - N_0) \quad (5.45)$$

Similar expressions can be found for the boundary conditions of the higher-order differential equations.

It must seem to the reader that the work required has been increased by carrying the intermediate quantity τ all through the expansions and then eliminating it at the last step, instead of eliminating it at once from Eqs. 5.1, 5.25, 5.29, and 5.33. The procedure was adopted to prepare for the work of the next section, where its usefulness will become apparent.

With this cataloguing of the boundary conditions all the development that can be made without special assumptions has been completed. The treatment has been more detailed than the novelty of the content strictly merits; it is hoped that its reference value will make up for the difference.

2. AVERAGE TRANSPORT LOSS FOR PERIODIC FLUCTUATIONS

There are two aspects of the effect of perturbations which are important. The first is to predict the concentration changes in detail, so that the success of corrective measures can be verified. The second is to find the loss of production over long periods of time.

If the primary disturbance $\sigma f(s,t)$ is nonperiodic, like a simple pulse or a damped oscillation, both aspects are of interest. The leading term in the answer is $N_1(s,t)$, but often $N_2(s,t)$ is also required to give a correct evaluation of the production loss.

If the primary disturbance is periodic, a new average steady state, with concentrations lower than before, will result. In this case we are more interested in the new steady state—the loss of production and average change in concentrations—than in the full detail of the answer. Usually[1] the leading term in the answer is the average second-order change in concentration or in transport. Of course the average second-order quantities can be found by solving straightforwardly for $N_2(s,t)$, as indicated in Sec. 1, and then averaging. But this procedure requires considerable work to get superfluous detail; it is possible to obtain[2] $\bar{\tau}_2$ and $\bar{N}_2(s)$ directly without solving the second-order differential equation. The method for doing this, which was devised by K. Fuchs, follows.

Suppose that the solution $N_1(s,t)$ or $N_1^*(s,p)$ of the equation of the first order in σ has been determined. Taking the average of Eq. 5.19 and using the relation Eq. 5.18 for D,

$$\bar{\tau}_2 = - c_{50} J^{-1} \frac{\partial(J\bar{N}_2)}{\partial s} + \bar{\Omega}_2(s) \qquad (5.46)$$

From Eq. 5.2, $\partial\bar{\tau}/\partial s = - \partial(c_6 N)/\partial t$. The right-hand side must be zero; otherwise there is no steady state. Therefore $\bar{\tau}$ is constant, except for a discontinuity at $s = 0$, and so are all orders of its expansion, including $\bar{\tau}_2$. The solution of the problem rests on this property of $\bar{\tau}_2$, which permits its evaluation from the boundary conditions.

Equation 5.46 integrates out to give, for the rectifier,

$$(J\bar{N}_2)_S - (J\bar{N}_2)_0 = - \bar{\tau}_2(S) \int_0^s \frac{J(s)\, ds}{c_{50}(s)} + \int_0^s \frac{\bar{\Omega}_2 J\, ds}{c_{50}} \qquad (5.47)$$

and, for the stripper (negative values of s),

$$(J'\bar{N}_2)_S - (J'\bar{N}_2)_0 = - \bar{\tau}_2(B) \int_0^s \frac{J'(s)}{c_{50}(s)}\, ds + \int_0^s \frac{\bar{\Omega}_2 J'}{c_{50}}\, ds \qquad (5.48)$$

[1]It may be supposed without loss of generality that $\sigma f(s,t)$ has a vanishing time average. It will then usually be true that the first-order terms in the fundamental quantities c_1, c_5, c_6, P ... have vanishing time averages:

$$\overline{c_{11}(s,t)} = \overline{c_{51}(s,t)} = \overline{c_{61}(s,t)} = \overline{P_1(t)} = \ldots = 0$$

Should this be the case, then $\overline{N_1(s,t)} = \bar{\tau}_1 = 0$, for, on averaging Eq. 5.10 and the boundary conditions (Eqs. 5.27, 5.31, and 5.35), it will be found that $\bar{N}_1$ satisfies an equation almost exactly like Eq. 5.9, but with the feed concentration zero instead of N_{F_0}. Note, however, that in the solution to be given for $\bar{N}_2$ and $\bar{\tau}_2$ we shall not assume that $\overline{c_{11}(s,t)}$, etc., and hence $\bar{N}_1$, equals zero.

[2]The time average of x(s,t) will be written as $\bar{x}(s)$ instead of the cumbersome $\overline{x(s,t)}$. The average of the product $x_1(s,t)x_2(s,t)$ will, however, be written as $\overline{x_1(s,t)x_2(s,t)}$.

It will be recalled that (Eq. 5.17)

$$J(s) = \exp\left[-\int_0^s \frac{c_{10}(1 - 2N_0) + P_0}{c_{50}} ds\right]$$

for the rectifier. J′ has been used to denote the corresponding function for the stripper.

$$J'(s) = \exp\left[-\int_0^s \frac{c_{10}(1 - 2N_0) - W_0}{c_{50}} ds\right] \qquad (5.49)$$

It now remains only to evaluate the as yet unknown constants $\overline{\tau}_2(S)$ and $\overline{\tau}_2(B)$ from the boundary conditions, and the problem is solved. From the boundary conditions of Eq. 5.28, at s = 0, on averaging,

$$\overline{\tau}_2(S) - \overline{\tau}_2(B) = F_0\overline{N}_{F_2} + \overline{F_1N_{F_1}} + \overline{F}_2N_{F_0} = A \qquad (5.50)$$

where A is a known constant. Also

$$\overline{N}_2(0+) = \overline{N}_2(0-) \qquad (5.51)$$

Now J(0) = J′(0) = 1. Thus with the aid of Eq. 5.51,

$$(J\overline{N}_2)_0 = (J'\overline{N}_2)_0$$

Setting s = S in Eq. 5.47 and s = B in Eq. 5.48 and subtracting,

$$(J\overline{N}_2)_S - (J'\overline{N}_2)_B = -\overline{\tau}_2(S)\int_0^S \frac{J(s)\,ds}{c_{50}(s)} + \overline{\tau}_2(B)\int_0^B \frac{J'(s)\,ds}{c_{50}(s)} + \int_B^S \frac{\overline{\Omega}_2 J\,ds}{c_{50}} \qquad (5.52)$$

It should be understood that J in the last integral is given by Eq. 5.49 from B to 0 and by Eq. 5.17 from 0 to S.

Now the boundary conditions (Eq. 5.32) at the top of the rectifier give a linear relation between $\overline{N}_2(S)$ and $\overline{\tau}_2(S)$; those at the bottom of the stripper (Eq. 5.36) give another between $\overline{N}_2(B)$ and $\overline{\tau}_2(B)$. Using these to eliminate $\overline{N}_2(S)$ and $\overline{N}_2(B)$ from Eq. 5.52, a linear relation between $\overline{\tau}_2(S)$ and $\overline{\tau}_2(B)$ is obtained, which, together with Eq. 5.52, allows the determination of both quantities.

These linear relations will be shown, but the process of eliminating $\overline{N}_2(S)$ and $\overline{N}_2(B)$ and solving for the $\overline{\tau}_2$ will not be undertaken here.

In the second equation of Eq. 5.32,

$$\frac{\overline{\partial N_{P_2}}}{\partial t} = \lim_{t \to \infty} \frac{N_{P_2}(t) - N_{P_2}(0)}{t} = 0$$

or else $N_{P_2}(t)$ increases or decreases indefinitely. Thus on averaging it is found that

$$c_{50}(P)[\overline{N}_2(S) - \overline{N}_{P_2}] = H_1 \overline{\frac{\partial N_{P_1}}{\partial t}} - \overline{c_{51}(P,t)\,[N_1(S,t) - N_{P_1}(t)]} \tag{5.53}$$

The first equation of Eq. 5.32 gives

$$\overline{\tau}_2(S) - P_0\overline{N}_{P_2} = \overline{N_{P_1}P_1(t)} + N_{P_0}\overline{P}_2 \tag{5.54}$$

Eliminating $\overline{N}_{P_2}$ between Eqs. 5.53 and 5.54 gives $\overline{N}_2(S)$ in terms of $\overline{\tau}_2(S)$, as required. Similarly, from the boundary condition (Eq. 5.36) at $s = B$,

$$c_{50}(W)\ [\overline{N}_2(B) - \overline{N}_{W_2}] = H_1' \overline{\frac{\partial N_{W_1}}{\partial t}} - \overline{c_{51}(W,t)\ [N_1(B,t) - N_{W_1}(t)]} \tag{5.55}$$

$$\overline{\tau}_2(B) + W_0\overline{N}_{W_2} = -\overline{W_1(t)N_{W_1}} - \overline{W}_2N_{W_0} \tag{5.56}$$

For many purposes $\overline{\tau}_2(S)$, which is the average (second-order) change in production of desired material, is a sufficient answer, and it is unnecessary to return to Eqs. 5.47 and 5.48 to find $\overline{N}_2(s)$.

The solution just given requires the averaging of some second-order quantities, and of products of first-order quantities. If these functions are known explicitly in terms of t, the averaging is done by elementary integration. However, they will usually be given as Laplace transforms. To perform the averaging without taking the inverse the following theorem,[1] which holds if a steady state exists, can be used: If $\lim px^* = C(s)$ as $p \to 0$ by positive real values, then $\overline{x}(s) = C(s)$; otherwise $\overline{x}(s) = 0$.

[1] G. Doetsch, "Theorie und Anwendung der Laplace Transformation," p. 193, Dover Publications, New York, 1943.

3. VARIATIONS IN PRODUCTION RATE FOR A SQUARE CASCADE

In this section the theory of the preceding sections will be illustrated and clarified by applying it to a special case. The abstractness of the demonstrations just given will be relieved by discussing further properties of the solutions with a particular example. The example chosen is that of a square cascade—practical comparison may be made with a fractionating column—on which are imposed arbitrary variations in P, the rate of production. Suppose that the hydrodynamics of the cascade are such that at the same time everything else is invariant: the holdups, the flows and separation factors, and the feed rate and composition. To simplify the arithmetic a rectifier alone will be considered, with the feed concentration held fixed. It is further supposed that N is always small. A reservoir is included at the top.

The terms c_{10}, c_{50}, and c_{60} do not vary with stage number. Let $c_{50}(P) = c_{50}$ (flow into reservoir = flow between stages). Introducing the same notation as in Chap. 3,

$$\lambda = \frac{c_{60}}{c_{50}}$$
$$2\epsilon = \frac{c_{10}}{c_{50}}$$
$$\psi = \frac{P}{c_{10}} \tag{5.57}$$
$$\epsilon' = \epsilon(1 + \psi)$$

The new constant ϵ' is used to condense the notation. The expansions of the coefficients reduce to

$$T(s,t) = P(t) = P_0 + \sigma P_1(t) \tag{5.58}$$

Everything else (except of course N and τ) has only zero-order terms. In place of the boundary conditions at s = 0 and s = B previously used, there is the simple expression

$$N(0,t) = N_F = \text{constant}$$

and hence

$$n_0(0) = N_F$$
$$N_1(0,t) = N_2(0,t) = N_1^*(0,p) = N_2^*(0,p) = 0 \tag{5.59}$$

The boundary condition at the top is still Eq. 5.29 and its descendants, notably Eq. 5.43.

The solution of the zero-order differential equation (Eq. 5.9) contains nothing new. As in Chap. 2, Sec. 1, it is found that

$$N_0(s) = \frac{\psi}{1+\psi} N_{P_0} + \left(N_F - \frac{\psi}{1+\psi} N_{P_0}\right) e^{2\epsilon' s} \tag{5.60}$$

The first-order differential equation (Eq. 5.16) requires the evaluation of $\partial\Omega_1^*/\partial s$. From Eq. 5.14, in the present case

$$\begin{aligned} \Omega_1 &= P_1 N_0 \\ \frac{\partial \Omega_1^*}{\partial s} &= P_1^* \frac{\partial N_0}{\partial s} = P_1^* N_F \frac{2\epsilon' e^{2\epsilon' s}}{1+\psi e^{2\epsilon' S}} \end{aligned} \tag{5.61}$$

and Eq. 5.16 becomes

$$\frac{\partial^2 N_1^*}{\partial s^2} - 2\epsilon' \frac{\partial N_1^*}{\partial s} - \lambda p N_1^* = \lambda C P_1^* e^{2\epsilon' s} \tag{5.62}$$

where

$$C = \frac{N_F 2\epsilon'}{\lambda c_{50}\left(1+\psi e^{2\epsilon' S}\right)} = \frac{N_{P_0} 2\epsilon e^{-2\epsilon' S}}{\lambda c_{50}} \tag{5.63}$$

The general solution of Eq. 5.62 is

$$N_1^* = e^{\epsilon' s}(A \cosh \beta s + B \sinh \beta s) - \frac{CP_1^*}{p} e^{2\epsilon' s} \tag{5.64}$$

As in Chap. 3,

$$\beta = \sqrt{\epsilon^2(1+\psi)^2 + \lambda p} = \sqrt{(\epsilon')^2 + \lambda p} \tag{5.65}$$

The constants A and B are evaluated from the boundary conditions (Eqs. 5.43 and 5.59). The latter in the present example reduces to

$$\left(2\epsilon' - \frac{\partial}{\partial s}\right) N_1^* = M(p) N_1^* \tag{5.66}$$

where

$$M(p) \equiv \frac{2\epsilon\psi + K_0 p}{1 + K_0 p} \tag{5.67}$$

Performing the indicated operations and collecting,

$$N_1^*(s,p) = \frac{CP_1^*}{p}\left[e^{\epsilon' s}\frac{(M-\epsilon')\sinh\beta(S-s)+\beta\cosh\beta(S-s)}{(M-\epsilon')\sinh\beta S+\beta\cosh\beta S} + e^{\epsilon' s}\frac{Me^{\epsilon' S}\sinh\beta s}{(M-\epsilon')\sinh\beta S+\beta\cosh\beta S} - e^{2\epsilon' s}\right]$$

$$= \frac{CP_1^*}{p}F(s,p) \qquad (5.68)$$

If the disturbance is taken to be sinusoidal, i.e.,

$$P_1 = \cos\omega t \qquad \text{and} \qquad P_1^* = \frac{p}{p^2+\omega^2} \qquad (5.69)$$

and then Eq. 5.68 is inverted, contributions are obtained from the poles at $\pm i\omega$, also from the roots of $(M-\epsilon')\sinh\beta S + \beta\cosh\beta S = 0$, which lie on the negative real axis. There is no pole at $p = 0$. Thus

$$N_1(s,t) = \frac{C}{2i\omega}F(s,i\omega)e^{i\omega t} - \frac{C}{2i\omega}F(s,-i\omega)e^{-i\omega t} + \sum_j A_j(s,p_j)e^{-|p_j|t} \qquad (5.70)$$

i.e., terms of the same period as the disturbance, plus transients. This is a general result. Note also that the form of Eq. 5.68, which is linear in P_1^*, allows superposition of solutions, i.e., a Fourier analysis of the variations in P, c_5, c_1, etc., is permissible.

In what follows a type disturbance more complicated than the sinusoidal will be discussed. Suppose that the product is withdrawn in batches at regular intervals a, instead of continuously. Then the disturbance function is

$$P_1(t) = -P_0 + \text{pulses including a quantity } aP_0 \text{ at } t = 0, a, 2a\ldots \qquad (5.71)$$

The transform of P_1 is easier to write

$$P_1^*(p) = \frac{-P_0}{p} + aP_0\sum_{k=0}^{\infty} e^{-kap} = \frac{-P_0}{p} + \frac{aP_0}{1-e^{-ap}} \qquad (5.72)$$

$P_1^*(p)$ has no pole at $p = 0$. The average of $N_1(s,t)$ is found from Eqs. 5.68 and 5.72:

$$\overline{N}_1(s) = \lim_{p\to 0} pN_1^*(s,p) = 0$$

(This is a particular example of the general result proved in Sec. 2.) To get the average loss of production, the second-order terms are thus necessary.

Equation 5.52 of Sec. 2 reduces now to

$$(J\overline{N}_2)_S = -\overline{\tau}_2 \int_0^S \frac{J(s)\, ds}{c_{50}} + \int_0^S \frac{\overline{\Omega}_2 J\, ds}{c_{50}} \tag{5.73}$$

and the boundary condition which determines $\overline{\tau}_2$ is, from Eqs. 5.53 and 5.54,

$$\overline{\tau}_2 - P_0\overline{N}_2(S) = \overline{P_1(t)N_{P_1}} \tag{5.74}$$

Now from Eq. 5.17,

$$J = \exp\left(-\int_0^s 2\epsilon'\, ds\right) = \exp(-2\epsilon' s) \tag{5.75}$$

Combining Eqs. 5.73, 5.74, and 5.75 and simplifying the result with the aid of Eq. 5.60, then

$$\frac{\overline{\tau}_2}{P_0 N_{P_0}} = \frac{\overline{P_1(t)N_{P_1}}}{P_0 N_F} + \frac{e^{2\epsilon' S}}{N_F c_{50}} \int_0^S \overline{\Omega}_2 e^{-2\epsilon' s}\, ds \tag{5.76}$$

Accordingly it is necessary to evaluate the expressions on the right-hand side of Eq. 5.76. From Eq. 5.20, $\Omega_2 = N_1 P_1$,

$$\begin{aligned} \Omega_2^*(s,p) &= \frac{1}{2\pi i} \int_{Br_1} N_1^*(s,p-q) P_1^*(q)\, dq \\ &= \frac{C}{2\pi i} \int_{Br_1} F(s,p-q) \frac{P_1^*(p-q) P_1^*(q)}{p-q}\, dq \end{aligned} \tag{5.77}$$

This integral is equal[1] to the sum of the residues at the poles of $P_1^*(q)$. These are at $q = 2k\pi i/a$ $(k = \pm 1, \pm 2 \ldots)$. Each residue has the value

$$\frac{CP_0}{p - \dfrac{2k\pi i}{a}} \left(-\frac{P_0}{p - \dfrac{2k\pi i}{a}} + \frac{aP_0}{1 - e^{-ap}} \right) F\left(s, p - \frac{2k\pi i}{a}\right)$$

[1]Strictly speaking

$$\Omega_2^* = \lim_{\nu \to \infty} + \sum_{k=-\nu}^{\nu} (\text{residues})$$

Then

$$\bar{\Omega}_2(s) = \lim_{p\to 0} p\Omega_2^* = CP_0^2 \sum_{-\infty}^{+\infty} \frac{a}{2k\pi i} F\left(s, \frac{2k\pi i}{a}\right) \tag{5.78}$$

and likewise, using the relation (Eq. 5.41) for $N_{P_1}^*$,

$$\overline{P_1(t)N_{P_1}} = CP_0^2 \sum_{-\infty}^{+\infty} \frac{a}{2k\pi i} \frac{F\left(s, \frac{2k\pi i}{a}\right)}{1 + \frac{K_0 2k\pi i}{a}} \tag{5.79}$$

The meaning of Eq. 5.78 will be clear by returning to the disturbance $P_1 = \cos \omega t$ and following through the same procedure. From Eq. 5.70 and $\Omega_2 = N_1 P_1$, this time averaging by direct integration with respect to t,

$$\bar{\Omega}_2 = \frac{\omega}{2\pi}\int_0^{2\pi/\omega}\left[\frac{C}{2i\omega} F(s,i\omega)e^{i\omega t} - \frac{C}{2i\omega} F(s,-i\omega)e^{-i\omega t}\right] \cos \omega t \, dt$$

$$\bar{\Omega}_2(s) = \frac{C}{4i\omega} F(s,i\omega) - \frac{C}{4i\omega} F(s,-i\omega) \tag{5.80}$$

Equation 5.80 is the contribution to $\bar{\Omega}_2(s)$ of a sinusoidal variation of frequency $\omega/2\pi$. Now, if the (improper) disturbance function (Eq. 5.71) is analyzed, representing batch withdrawal as

$$P_1(t) = 2P_0 \sum_{k=1}^{\infty} \cos \frac{2k\pi t}{a}$$

and if contributions of the individual frequencies given by Eq. 5.80 are summed, the result is precisely Eq. 5.78.

This is a general result, which can be understood as follows: In a disturbance, two different frequencies (in P_1, for example) contribute independently to N_1, each giving a term of its own frequency. In taking time averages, terms like $e^{\pm\omega_1 t \pm \omega_2 t}$ always vanish. The result is a sum of contributions from each frequency acting as if it were alone. This whole process is taken care of automatically by the use of the averaging theorem.[1]

[1] In comparing the solution of the differential equation by the functionally closely related methods of the Laplace transform or the Fourier expansion, it should be remarked that the former method gives at the same time the solution for the transients.

In the same way the effect of disturbances of the same frequency which are out of phase by an angle α can be realized. Corresponding to Eq. 5.70, N_1 is

$$N_1(s,t) = \frac{C}{i\omega} F(s,i\omega)e^{i\omega t} + \frac{C}{i\omega} F(s, i\omega)e^{i(\omega t+\alpha)} + \ldots$$

and on calculating $\overline{N_1P_1}$ (or similar terms) there will result the contributions of the separate disturbances plus a cross-product term with cos α as a factor. Now if the two disturbances are unrelated cos α is, on the average, zero, and the resultant is purely additive. Note, however, that fluctuations in P and, for example, c_5 caused by a common primary disturbance are correlated through the hydrodynamic response of the plant and cannot be treated separately.

The remarks of the preceding two paragraphs contain the results necessary for the composition of disturbances of different sources and types.

Returning to the derivation of $\bar{\tau}_2$ according to Eq. 5.76 and picking up the thread of the argument at the expressions Eqs. 5.78 and 5.80, consider $(1/i\omega)[F(s,i\omega) - F(s, -i\omega)]$ as a function of ω. Term F(s,p) is defined by Eq. 5.68. The term $e^{2\epsilon' s}$ does not depend on ω and hence gives nothing. Then

$$F \approx \frac{(M - \epsilon') \sinh \beta(S - s) + \beta \cosh \beta(S - s) + Me^{\epsilon' S} \sinh \beta s}{(M - \epsilon') \sinh \beta S + \beta \cosh \beta S} e^{\epsilon' S} \qquad (5.81)$$

where

$$\beta = \sqrt{\epsilon^2(1 + \psi)^2 \pm i\omega\lambda}$$

$$M = \frac{2\epsilon\psi \pm i\omega k_0}{1 \pm i\omega k_0}$$

For $\omega\lambda \gg \epsilon^2(1 + \psi)^2$, a condition fulfilled by all but the lowest frequencies because of the smallness of ϵ,

$$\beta \to \sqrt{\omega\lambda}\,(1 \pm i)$$

$$\sinh \beta(S - s) \to \cosh \beta(S - s) \to \tfrac{1}{2} e^{\sqrt{\omega\lambda}\,(S-s)(1\pm i)}$$

$$F(s, \pm i\omega) \to e^{-s\sqrt{\omega\lambda}\,(1\pm i)}$$

Then

$$\frac{1}{i\omega} [F(s,i\omega) - F(s, -i\omega)] \approx -\frac{2}{\omega} e^{-s\sqrt{\omega\lambda}} \sin s\sqrt{\omega\lambda} \qquad (5.82)$$

Now

$$\int_0^s \frac{1}{i\omega}\,[F(s,i\omega) - F(s,-i\omega)]\,e^{-2\epsilon' s}\,ds \approx -\frac{1}{\sqrt{\omega^3\lambda}} \tag{5.83}$$

The contribution to $\overline{\tau}_2/P_0N_{P_0}$ from the disturbance $P_1 = \cos\omega t$ is, by Eq. 5.76,

$$\frac{e^{2\epsilon' s}}{N_F c_{50}} \int_0^s \overline{\Omega}_2 e^{-2\epsilon' s}\,ds = -\frac{\epsilon}{2c_{50}^2}\,\frac{N_{P_0}}{N_F}\,\frac{1}{\sqrt{\omega^3\lambda^3}} \tag{5.84}$$

The integral was evaluated by Eqs. 5.80 and 5.83 and by the definition of C, Eq. 5.63. The contribution from the term $\overline{P_1(t)N_{P_1}}/P_0$ vanishes in the same manner as $e^{-s\sqrt{\omega\lambda}}/\omega$ and can be neglected.

The disturbance in Eq. 5.71 caused by batch withdrawal gives likewise, comparing Eq. 5.78 with Eq. 5.80,

$$\begin{aligned}\frac{\overline{\tau}_2}{PN_{P_0}} &= -\left(\frac{a}{2\pi\lambda}\right)^{3/2} 2\epsilon\,\frac{P_0^2}{c_{50}^2}\,\frac{NP_0}{N_F}\sum_{k=1}^{\infty} k^{-3/2} \\ &= -\,20.9\left(\frac{a\epsilon^2}{2\pi\lambda}\right)^{3/2} \psi^2\,\frac{N_{P_0}}{N_F}\end{aligned} \tag{5.85}$$

provided $a \ll 2\pi\lambda/\epsilon^2(1+\psi)^2$. The holdup in the reservoir has no influence to this degree of approximation. For larger values of a, numerical summations are needed.

The factors in the solution (Eq. 5.85) are a term in the ratio of a to the natural period, λ/ϵ^2; a factor ψ^2 that is related to the size of the disturbance; and a factor N_{P_0}/N_F that shows that the sensitivity is proportional to the enrichment factor of the column. It is interesting to note that the period λ/ϵ^2 and not $-1/p_1$, the relaxation time, enters in; also that the magnitude of disturbance P_0 enters in as $\psi = P_0/c_{10}$ and not P_0/c_{50}. An idea of the orders of magnitude involved is given in Table 5.1, which compares two cases.

In both cases a is about equal to the time for the column to reach the enrichment N_{P_0}/N_F. However, in case A, $a\epsilon^2/\lambda = 1/40$, whereas in case B $a\epsilon^2/\lambda = 1$, so that the loss in case B is much more serious. The second most important factor is the severity of the disturbance; $\psi = 2$ in case A and $\psi = 0.2$ in case B. The weight of this factor is in the other direction. By comparison the differences in the factor N_{P_0}/N_F are slight.

In case B, $\bar{\tau}_2$ is already quite large; in fact it may be suspected that higher-order terms are needed. In case A, $\bar{\tau}_2$ becomes large if a is taken much larger, even though $a\epsilon^2/\lambda$ is still small. Thus in both cases $\bar{\tau}_2$ becomes appreciable at $a \approx -1/p_1$, although p_1 does not appear in the formula (Eq. 5.85). This can be understood after an examination of the transient terms in $N_1(s,t)$. The first transient has the factor

$$f = -\frac{1}{p_1} + \frac{a}{1 - e^{-ap_1}}$$

This factor increases rapidly for fixed p_1 as ap_1 becomes large. For example, taking $p_1 = -1$: for $a = 0$, $f = 0$; for $a = 0.1$, $f = 0.048$; for $a = 1$, $f = 0.42$; for $a = 10$, $f = 100$. Thus the transients get large as $a \sim -1/p_1$, implying that the new steady state differs significantly from the initial state. Holdup at the top increases $-1/p_1$ and must therefore permit larger values of a, even though this is not exhibited by the rough formula, Eq. 5.85.

Table 5.1 — Production Loss Due to Batch Withdrawal

	Case A	Case B
Equilibrium enrichment (no production)	2-fold	30-fold
$2\epsilon S$	0.693	3.40
λ/ϵ^2	40 days	1 day
First transient period:[1]		
$-1/p_1$	1.16 days	0.96 day
ψ	2	0.2
N_{P_0}/N_F	$\sqrt{2}$	$\sqrt{30}$
a	1 day	1 day
$\bar{\tau}_2/P_0N_{P_0}$	−3.5 %	−29%

[1]Taking the holdup in the top reservoir as zero.

The evaluation of the loss in production by the single term $\bar{\tau}_2$ will be correct only if the higher-order terms are negligible, i.e., only if $\bar{\tau}_2$ is quite small. This slightly discouraging conclusion does not deprive the calculation of its usefulness, which is mainly to assign permissible upper limits to the disturbances.

It is difficult to generalize from the problems just solved to other problems. Even the most interesting result, that the effect of high-frequency disturbances vanishes as $\omega^{-3/2}$, is only true for a disturbance in P alone. It does not subsist when fluctuations in the coefficients c_6, c_5, and c_1 are taken into consideration. The correlations to be expected between disturbances of the various coefficients are different from one separating apparatus to another. Hence this is about as far as the investigation can proceed without specifying the type of equipment in question.

The reader may have noted that, leaving out the explanations and digressions, the solution to the problem of this section was obtained fairly easily. This is not generally true; problems on more intricate cascade shapes and for intermediate frequencies run into heavy computation. However, it should be remembered, in spite of the space required to expose the theory, that small correction terms are involved. It is therefore quite proper to reduce the numerical work by rather drastic approximations.

REFERENCES

Fuchs, K., British MS 71, January–November, 1943.

Fuchs, K., British Supply Council reports: MSN 4, Feb. 11, 1944; MSN 5, Mar. 4, 1944; MSN 8, Mar. 7 and Mar. 17, 1944; MSN 10, Apr. 6, 1944; MSN 12, June 6, 1944; MSN 14, June 28, 1944; MSN 15, July 11, 1944; MSN 17, July 20, 1944.

Chapter 6

CENTRIFUGES

1. HISTORICAL INTRODUCTION

The effect of the earth's gravitational field on the composition of the gases in the atmosphere at various heights has long been recognized. For, if p_i is the partial pressure of a gas of molecular weight M_i, according to the Boltzmann distribution law

$$p_i(h) = p_i(0) \exp - \frac{M_i gh}{RT} \qquad (6.1)$$

where h is the height; g is the gravitational constant; R is the gas constant, 8.314×10^7 ergs/°K/mole; and T is the absolute temperature. Then for two gases of different molecular weights M_1 and M_2

$$\frac{p_1(h)}{p_2(h)} = \frac{p_1(0)}{p_2(0)} \exp \frac{(M_2 - M_1)gh}{RT} \qquad (6.2)$$

The suggestion that gravitational or centrifugal fields are suitable for isotope separation was made by Lindemann and Aston[1] in 1919. In fact, they proposed that samples of neon taken at ground level and at 100,000 ft would show a sensible shift in the isotopic abundance ratio.

In a centrifuge, vastly more powerful fields are available. The energy per mole at radius r in a centrifuge rotating with angular velocity ω is $\frac{1}{2} M_i(\omega r)^2$. This expression for the energy replaces $-M_i gh$ in Eq. 6.1, giving

$$p_i(r) = p_i(0) \exp \frac{M_i(\omega r)^2}{2RT} \qquad (6.3)$$

[1] Phil. Mag., 37: 530 (1919).

Between the periphery $r = r_2$ and axis $r = 0$, for two different gases, from Eq. 6.3,

$$\frac{p_1(r_2)}{p_2(r_2)} = \frac{p_1(0)}{p_2(0)} \exp \frac{(M_1 - M_2)(\omega r_2)^2}{2RT} \tag{6.4}$$

where ωr_2 is the velocity of the inside wall of the rotor, the inner peripheral velocity.

In a binary system, if N is the mole fraction of the isotope having mass M_1,

$$\begin{aligned} p_1 + p_2 &= p \\ N &= p_1/p \\ 1 - N &= p_2/p \end{aligned} \tag{6.5}$$

and $p - p(r)$ is the total pressure. Then Eq. 6.4 becomes

$$\left(\frac{N}{1-N}\right)_{r_2} = \left(\frac{N}{1-N}\right)_0 \exp \frac{(M_1 - M_2)(\omega r_2)^2}{2RT} \tag{6.6}$$

It is usual to take $M_2 > M_1$, and accordingly

$$\alpha_0 = \exp \frac{(M_2 - M_1)(\omega r_2)^2}{2RT} \tag{6.7}$$

so that Eq. 6.6 takes the form

$$R(0) = \alpha_0 R(r_2)$$

where α_0 is the (equilibrium) simple-process factor of the separation.

The value of the simple-process factor for a peripheral velocity of 300 meters/sec, $T = 300°K$, and a mass difference of one unit is, by Eq. 6.6, $\alpha_0 = 1.0183$.

Shortly after the suggestion of Lindemann and Aston was made, attempts were made, in rather crude apparatus, by Joly and Poole[1] and by Mulliken[2] to realize this separation factor, but these were unsuccessful. The idea of the evaporative centrifuge was introduced at the same time by Mulliken. In this method a small amount of liquid is introduced into the centrifuge, forming a layer at the periphery. Dur-

[1] Phil. Mag., 39: 372 (1920).
[2] J. Am. Chem. Soc., 44: 1033 (1922).

ing the spinning of the rotor, vapor is removed slowly through a shaft along the axis. In this way a Rayleigh distillation (Appendix D) of the liquid is accomplished, with simple-process factor given by Eq. 6.31. An increased concentration change in the residual liquid is thus obtained.

The development by Beams[1] of vacuum-chamber centrifuges, which were free from vibration and thermally isolated to eliminate convection currents, finally permitted successful centrifugal isotope separation to be obtained in 1939. Beams and Skarstrom[2] at the University of Virginia, using the evaporative centrifuge method on CCl_4, reported a 13 per cent change in the Cl^{35}/Cl^{37} ratio. Shortly thereafter Humphreys,[3] using the same technique on ethyl bromide, altered the Br^{79}/Br^{81} abundance ratio by 11 per cent.

The simple-process factor of centrifugal separation depends only on the mass difference. Unlike other methods (e.g., diffusion through barriers, where α varies as $\Delta M/M$) it is no more difficult for heavy elements than for light. In addition, it is possible to calculate a priori the separation to be expected from centrifugal apparatus without any undetermined constants or theoretical uncertainties.

The simple-process "flow-through centrifuge," as the concurrent flow type was called, is an application to gaseous flow of the continuous cream separator. As originally conceived, a single stream of gas enters one end of a rotor through a hollow shaft, and two streams are taken off the other end, one from the periphery and the other near the axis. This method produces a small change in concentration per machine.

The countercurrent flow type was designed to attain considerable separation in a single centrifuge, thus reducing the number of stages required and the amount of material circulated between stages. As originally proposed by Urey,[4] circulation was to be established by continuous distillation of UF_6 from the bottom cap of the rotor. The vapor was to be condensed on the top cap. The heavy liquid would then be forced out to the periphery and would flow down the walls to the bottom cap, countercurrent to the vapor flow, and complete the cycle. Bramley and Brewer[5] and Martin and Kuhn[6] in Germany proposed to establish countercurrent flow by thermal convection.

[1] Revs. Modern Phys., 10: 245 (1938); Rev. Sci. Instruments, 9: 413 (1938).
[2] Phys. Rev., 56: 266 (1939).
[3] Phys. Rev., 56: 684 (1939).
[4] Reports on Progress in Physics, VI: 72 (1939).
[5] Science, 92: 427 (1940).
[6] Z. physik. Chem., A189: 219 (1940).

The aim of this chapter is to expose the theory of centrifugal separation for various types of centrifuges. The important question of mechanical design will be dealt with only in so far as it is essential to the understanding of the process design. The partial differential equation of centrifugal separation is derived in Sec. 2. In Sec. 3 the theoretical maximum separative power of a centrifuge is found. The solution for the evaporative centrifuge is given in Sec. 4, that for the concurrent centrifuge in Sec. 5, and that for the countercurrent centrifuge in Sec. 6. Section 6 investigates at length the properties of the countercurrent centrifuge.

The writer was introduced to the art of centrifugation by C. Skarstrom, formerly of the University of Virginia and later with the Standard Oil Development Company. Harold C. Urey participated actively in the early work, until his other responsibilities became too heavy, contributing ideas and physical interpretations of mathematical results. Irving Kaplan shared in all the developments subsequent to December 1941, notably the work on cascades of centrifuges (incorporated largely in Chaps. 1 to 5) and the design problem.

2. PARTIAL DIFFERENTIAL EQUATION OF CENTRIFUGE

As a preliminary to the derivation of the differential equation, consider for a moment a centrifuge containing a single pure gas. The gas is supposed to rotate uniformly at the same speed as the rotor, i.e., ω radians/sec; to have no other motions; and to be isothermal. Choose a cylindrical coordinate system, rotating with the gas, with the origin at the axes of rotation. The effect of the rotation is to set up a centrifugal force $\omega^2 r$ per unit mass. Under the influence of this force, a pressure gradient is established according to the hydrodynamic equation

$$\frac{1}{\rho}\frac{\partial p}{\partial r} = \omega^2 r \qquad (6.8)$$

Here ρ is the density of the gas mixture in grams per cubic centimeter. The quantities p and ρ are related by the gas law

$$p = \frac{RT\rho}{M} \qquad (6.9)$$

Combining Eqs. 6.8 and 6.9 gives Eq. 6.3 or its equivalent

$$\rho(r) = \rho(0) \exp \frac{M\omega^2 r^2}{2RT} \qquad (6.10)$$

The density distribution (Eq. 6.10), considered from a molecular point of view, is a dynamic equilibrium between the effect of random motion of the molecules (diffusion), which tends to erase density gradients, and the effect of the centrifugal field, which tries to pile all the molecules at the periphery. The diffusion counter to the density gradient gives a current density (in grams per second per square centimeter)

$$-D\,\frac{\partial \rho}{\partial r} = -D\rho\,\frac{M\omega^2 r}{RT}$$

Here D is the coefficient of self-diffusion. The outward current per square centimeter caused by the centrifugal force, which balances this inward current, must therefore be

$$\rho u_c = +D\rho\,\frac{M\omega^2 r}{RT} \qquad (6.11)$$

In mixtures, Eq. 6.11 holds for each component separately

$$\rho_i u_{c_i} = D_i\,\rho_i\,\frac{M_i\omega^2 r}{RT} \qquad (6.12)$$

D_i is the diffusion coefficient for molecules of type i through the whole mixture. For isotopic mixtures the exact meaning of the diffusion coefficient is of no significance, and henceforth the subscript will be dropped. Equation 6.12 is the desired expression for the flux of molecules caused by a centrifugal field, and it is now possible to derive the differential equation.

Consider then a centrifuge containing a perfect binary gas mixture, isothermal and rotating uniformly. In order to accomplish separation, the gas is made to flow radially and axially. In terms of the cylindrical coordinate system chosen as before, the motion of the gas as a whole (drift velocity) is given by the components

(Radial)	$u = u(r,z)$
(Azimuthal)	$v = 0$
(Axial)	$w = w(r,z)$

Because of this motion of the gas, the equilibrium distribution (Eq. 6.3 or 6.10) is continually upset. The system moves material in an effort to reestablish equilibrium; in this way a transport of the desired material is set up.

Imagine a small annular region in the rotor, bounded by the cylinders $r = r_0$ and $r = r_0 + \delta r$ and by the planes $z = z_0$ and $z = z_0 + \delta z$. In a short time interval δt there flows across the cylindrical face $r = r_0$ an amount of desired isotope (denoted by the subscript 1),

$$g_1(r_0) = \left(\rho_1 u + \rho_1 u_{c_1} - D\,\frac{\partial \rho_1}{\partial r}\right) 2\pi r_0\, \delta z\, \delta t \tag{6.13}$$

In this equation $\rho_1 u$ is the flow caused by the drift velocity, $\rho_1 u_{c_1}$ is the flow impressed by the centrifugal field, $-D\,\partial\rho_1/\partial r$ is the contribution from diffusion against the density gradient, and $2\pi r_0\,\delta z$ is the area of the cylindrical face. Likewise the flow across the plane face $z = z_0$ is

$$g_1'(z_0) = \left(\rho_1 w - D\,\frac{\partial \rho_1}{\partial z}\right) 2\pi r_0\, \delta r\, \delta t \tag{6.14}$$

The change in content of desired material in the volume element is therefore

$$\begin{aligned} \frac{\partial \rho_1}{\partial t}\, 2\pi r_0\, \delta r\, \delta z\, \delta t &= g_1(r_0) - g_1(r_0 + \delta r) - g_1'(z_0) - g_1'(z_0 + \delta z) \\ &= -\frac{\partial g_1}{\partial r}\,\delta r - \frac{\partial g_1'}{\partial z}\,\delta z \end{aligned} \tag{6.15}$$

Introducing Eqs. 6.13 and 6.14 into 6.15 and making use of the relation of Eq. 6.12 for u_{c_1} (dropping the subscript 0),

$$\frac{\partial \rho_1}{\partial t} = -\frac{1}{r}\,\frac{\partial}{\partial r}\, r\left(D\rho_1\,\frac{M_1\omega^2 r}{RT} - D\,\frac{\partial \rho_1}{\partial r} + \rho_1 u\right) - \frac{\partial(\rho_1 w)}{\partial z} + D\,\frac{\partial^2 \rho_1}{\partial z^2} \tag{6.16}$$

Now

$$\rho_1 = \frac{M_1}{RT}\, p_1 = \frac{M_1}{RT}\, Np \tag{6.17}$$

and

$$\frac{\partial p}{\partial r} = \omega^2 r(\rho_1 + \rho_2) = \frac{\omega^2 r}{RT}\, p[M_1 N + M_2(1 - N)] \tag{6.18}$$

Substituting Eq. 6.17 in Eq. 6.16, using Eq. 6.18 and noticing that Dp = constant, it is found that

$$p\frac{\partial N}{\partial t} = \frac{Dp}{r}\frac{\partial}{\partial r}\left[r\frac{\partial N}{\partial r} + \frac{(M_2 - M_1)\omega^2 r^2}{RT}N(1-N)\right]$$
$$-\frac{1}{r}\frac{\partial}{\partial r}(Npru) - \frac{\partial}{\partial z}(Npw) + D\frac{\partial^2}{\partial z^2}(Np) \qquad (6.19)$$

which is the desired partial differential equation of centrifugal separation.

The three kinds of centrifuge described in Sec. 1 are governed by special cases of Eq. 6.19. In the evaporative centrifuge $w = 0$ and ru = constant; in the concurrent and countercurrent centrifuges $u = 0$, and w describes the respective types of flow. The solution of Eq. 6.19 for these cases is the subject of the subsequent sections devoted to the particular centrifuge types.

3. MAXIMUM SEPARATIVE POWER OF A CENTRIFUGE

In this section the maximum separative work that can be performed by a centrifuge[1] will be determined. It will be used as a standard to evaluate the efficiency of the various centrifuge designs.

Using the result of Sec. 6 (Eq. 1.60),

$$\delta U = \frac{(\tau - NP)(N' - N)}{N^2(1-N)^2}$$

The net transport of desired material (in moles per second) across a surface element $d\sigma$ is, according to the analysis given in a preceding section,

$$\frac{d\sigma}{M_1}\left(\rho_1\vec{u}_{c_1} - D\nabla\rho_1\right) = \frac{Dp}{RT}\left[(M_1 - M_2)\frac{\omega^2 r}{RT}N(1-N)\vec{r}_0 - \nabla N\right]d\sigma \qquad (6.20)$$

where $\vec{r}_0$ is a unit vector along a radius through $d\sigma$, $\vec{u}_{c_1}$ is the vector velocity created by the centrifugal field directed along $\vec{r}_0$ and with magnitude given by Eq. 6.12, and the symbol ∇ is the vector gradient operator.

Thus the separative power for a small slab of thickness ds, with faces $d\sigma$ normal to ∇N and sides parallel to ∇N, is

$$dU = \frac{Dp}{RT}\,\frac{\left[(M_1 - M_2)\frac{\omega^2 r}{RT}N(1-N)\vec{r}_0 - \nabla N\right]\cdot\nabla N}{N^2(1-N)^2}\,d\sigma\,ds \qquad (6.21)$$

[1] Following Dirac.

The symbol · indicates the scalar product.

Taking ∇N as a variable and $d\sigma$ ds and everything else constant, Eq. 6.21 is a maximum when

$$\nabla N = \frac{1}{2}(M_1 - M_2)\frac{\omega^2 r}{RT} N(1-N)\bar{r}_0 \qquad (6.22)$$

that is, along the direction of the centrifugal field, and one-half the equilibrium value. Then

$$(dU)_{max} = \frac{Dp}{RT}\left[\frac{(M_2 - M_1)\omega^2 r}{2RT}\right]^2 d\sigma\, dr$$

Integrating over the whole centrifuge, of length Z and radius r_2,

$$(\delta U)_{max} = \int (dU)_{max} = \frac{Dp}{RT}\left[\frac{(M_2 - M_1)\omega^2}{2RT}\right]^2 \int_0^{r_2} r^2\, 2\pi r\, Z\, dr$$

$$= \frac{Dp}{RT}\left[\frac{(M_2 - M_1)\,(\omega r_2)^2}{2RT}\right]^2 \frac{\pi Z}{2} \quad \text{(moles/sec)} \qquad (6.23)$$

$$= \rho D\left[\frac{(M_2 - M_1)\,(\omega r_2)^2}{2RT}\right]^2 \frac{\pi Z}{2} \quad \text{(grams/sec)}$$

Note that the expression in brackets is the exponent of α_0 (Eq. 6.7).

The minimum number of centrifuges of length Z and peripheral velocity ωr_2 required to perform a given separation is, by Eqs. 1.44 and 6.23, $\Delta U/(\delta U)_{max}$. The value of $(\delta U)_{max}$ depends on the peripheral velocity of the fourth power, which puts a premium on high peripheral velocities. The separative power is also proportional to the length. But for constant peripheral velocity,[1] the separation is in-

[1]The peripheral velocity—not the angular velocity—is determined by the tensile strength of the rotor wall, so that it is the fundamental characteristic of the machine. The demonstration follows: Consider the equilibrium of forces acting on a sector of the rotor wall, subtending a small angle ϕ. Then the triangle of forces gives (Fig. 6.1)

$$\rho\, r\phi\, d\sigma\, \omega^2 r = 2T\, d\sigma \sin\frac{\phi}{2} = T\, d\sigma\, \phi$$

or

$$\omega^2 r^2 = \frac{T}{\rho}$$

where T is the stress in the wall and ρ is its density. The rotor will break when $T = \rho\omega^2 r^2$ is greater than the tensile strength of a metal.

dependent of the radius. The radius affects the holdup, and in counter-current centrifuges the flow into the rotors, but is otherwise irrelevant. It is usually chosen by mechanical considerations.

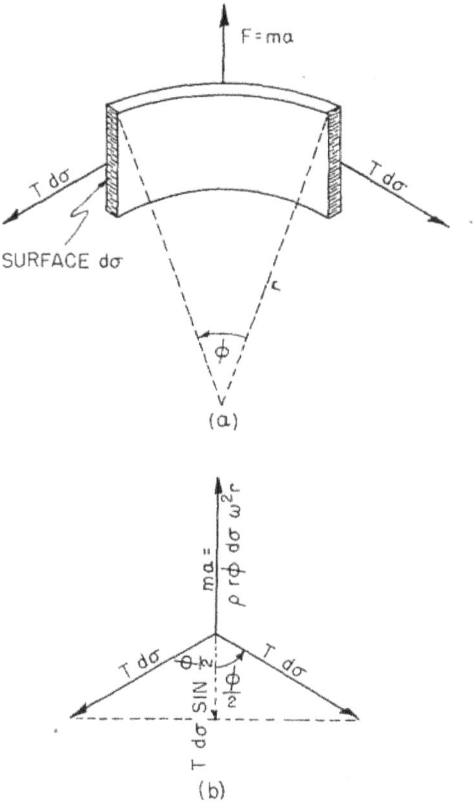

Fig. 6.1—Stresses in a thin rotating cylinder. (a) Stresses in a wall sector. (b) Vector diagram.

4. THE EVAPORATIVE CENTRIFUGE

In the evaporative centrifuge N is independent of z, and w = 0. Considering the steady state, $\partial N/\partial t = 0$ and Eq. 6.19 reduces to

$$\frac{1}{r}\frac{\partial}{\partial r}\left\{Dp\left[r\frac{\partial N}{\partial r} + 2\Delta A r^2 N(1-N)\right] - Npru\right\} = 0 \qquad (6.24)$$

where

$$\Delta A = \frac{M_2 - M_1}{2RT}\,\omega_2 \qquad (6.25)$$

Suppose vapor is removed from the axis at a rate L moles/sec. Then

$$-\frac{p}{RT}\,2\pi rZu = L \tag{6.26}$$

Multiplying Eq. 6.24 by the constant $2\pi Z/RT$ and integrating,

$$\gamma\left[r\,\frac{\partial N}{\partial r} + 2\Delta A r^2 N(1-N)\right] = L\,N(0) - L\,N(r) \tag{6.27}$$

where

$$\gamma = 2\pi Z\,\frac{Dp}{RT} \quad \text{(moles/sec)} \tag{6.28}$$

The integration constant was evaluated by noting that the left side of Eq. 6.27 vanishes at $r = 0$.

Replacing N by R as dependent variable,

$$\gamma\left(r\,\frac{\partial R}{\partial r} + 2\Delta A r^2 R\right) = L\left[R(0) - R(r)\right]\left[1 + \frac{R - R(0)}{1 + R(0)}\right] \tag{6.29}$$

Now $\frac{R - R(0)}{1 + R(0)} \leq -\Delta A r^2 N(0)$, which can be neglected with respect to unity. After thus simplifying Eq. 6.29, the solution is readily found to be

$$R = R(0)\exp\left[-\Delta A r^2\left(\frac{2\gamma}{2\gamma + L}\right)\right]$$

and on introducing Eq. 6.7, which in terms of ΔA is

$$\alpha_0 = \exp\left[+\Delta A r_2^2\right] \tag{6.30}$$

this expression may be written

$$R(0) = R(r_2)\,\alpha_0^{\left(\frac{2\gamma}{2\gamma+L}\right)} \tag{6.31}$$

Equation 6.31 gives the relation for mole fraction of product N(0) in terms of the composition of the charge $N(r_2)$ and the rate of production.[1] If it is assumed that the production is only a small fraction of the charge, or that successive batches of product are kept separate, the separative power of the centrifuge is (cf. Appendix D)

$$\delta U = \frac{1}{2} L \left(\alpha_0^{\left(\frac{2\gamma}{2\gamma+L}\right)} - 1 \right)^2 = \frac{1}{2} L \left(\frac{2\gamma}{2\gamma + L} \right)^2 (\ln \alpha_0)^2 \qquad (6.32)$$

The value of δU is a maximum when $L = 2\gamma$:

$$(\delta U)_{max} = \frac{\gamma}{4} (\ln \alpha_0)^2 = \frac{Dp}{RT} (\ln \alpha_0)^2 \frac{\pi Z}{2} \qquad (6.33)$$

which is the same as the theoretical maximum separative power of a centrifuge (Eq. 6.23).

In the development just given the gas was assumed to be isothermal. This picture is far too simple. When vapor is withdrawn at the axis, the gas cools by expansion and a temperature gradient is induced. The gradient is determined by the balance between the rate of cooling of the gas and thermal conductivity. As L increases, the gradient increases until the gas reaches the dew point or the adiabatic gradient. The mean effective temperature T_{eff} is therefore reduced as a function of L.

An attempt may be made to account for this effect by writing in place of Eq. 6.31

$$R(0) = R(r_2) \alpha_0^{\left(\frac{T}{T_{eff}} \frac{2\gamma}{2\gamma+L}\right)} \qquad (6.34)$$

Since T/T_{eff} increases with L, the separation will not fall off quite so fast with L in practice as Eq. 6.31 would indicate.

A unit weight of gas rotating near the periphery with angular velocity ω has the angular momentum $r_2^2\omega$; at the axis it has none. When gas is moved from the periphery toward the axis, means must be provided to absorb this excess angular momentum, which otherwise tends to increase the angular velocity of the gas at smaller radii. This is accomplished by radial baffles[2] (looking down along the axis, a star) which prevent the gas from rotating at angular velocities different from that of the rotor.

[1] This result was found by Humphreys, Phys. Rev., 56: 684 (1939).
[2] Beams and Skarstrom, Phys. Rev., 56: 266 (1939).

Although the evaporative centrifuge, with $L = 2\gamma$, delivers 100 per cent of the theoretical separative power, it is not easily adapted to continuous operation.

5. THE CONCURRENT CENTRIFUGE

Like the evaporative centrifuge, the concurrent centrifuge is a simple-process machine since the enrichment factor per machine is limited to α_0.

In this device axial flow is used to obtain an axial as well as a radial concentration gradient. The gas enters (Fig. 6.2) in two streams at one end of the rotor and flows axially to the other end, where the streams are removed separately. During the passage through the centrifuge the streams tend to assume the (radial) equilibrium distribution (Eq. 6.6). The flow pattern adopted for the concurrent centrifuge consists of two thin cylindrical streams located at radii r_1 and r_2 (periphery) and flowing parallel to the z axis. (The optimum position for the inner stream is not at the axis, as will be seen subsequently.) There is no radial mass flow.

A rotor with a double entrance (so that entering streams of two different concentrations can be used) will be considered; previously the literature has covered only single-entrance machines. The double-entrance centrifuge is more flexible than the latter. Cascades of these elements, which are more general than the element of Chaps. 1 to 5, are studied in detail in Appendix E. Among the more interesting features of this study is the result that the no-mixing case (both entering streams of the same composition) is not an optimum.

The solution given here follows roughly the demonstration by Murphree, which for the case considered is superior to the more general method that was originally used. The value of u is zero, and the flows are generally so large that back-diffusion is negligible. Then Eq. 6.19 reduces for the steady state to

$$0 = \frac{Dp}{r}\frac{\partial}{\partial r}\left[r\frac{\partial N}{\partial r} + 2\Delta A r^2 N(1-N)\right] - \frac{\partial}{\partial z}(Npw) \qquad (6.35)$$

In the region between $r = r_1$ and $r = r_2$, $w = 0$, so that Eq. 6.35 integrates to

$$-\frac{Dp}{RT}2\pi\left[r\frac{\partial N}{\partial r} + 2\Delta A r^2 N(1-N)\right] = f(z) \qquad (6.36)$$

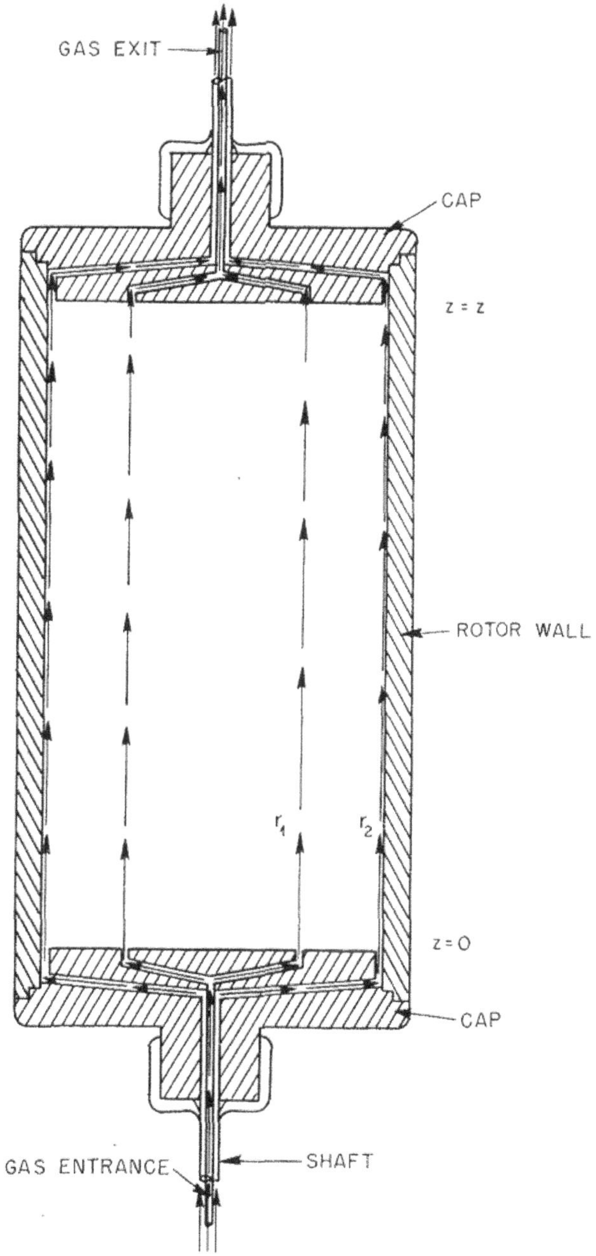

Fig. 6.2—Simple model of a gaseous concurrent centrifuge.

where $f(z)$ is the transport of desired isotope across a cylindrical element of unit length between the two streams.

But also, if $L(r_2)$ is the flow in the stream at $r = r_2$,

$$f(z) = L(r_2)\frac{d}{dz}N(r_2,z) = -L(r_1)\frac{d}{dz}N(r_1,z) \tag{6.37}$$

and

$$\begin{aligned} L(r_2)N(r_2,z) + L(r_1)N(r_1,z) &= L(r_2)N(r_2,0) + L(r_1)N(r_1,0) \\ &= B \quad \text{(a constant)} \end{aligned} \tag{6.38}$$

Integrating Eq. 6.36 between r_1 and r_2, considering $\Delta A\, N(1-N)$ constant to terms of the order of $(\Delta A)^2$,

$$-\frac{Dp}{RT}2\pi[N(r_2,z) - N(r_1,z) + \Delta A(r_2^2 - r_1^2)N(1-N)] = f(z)\ln\frac{r_2}{r_1} \tag{6.39}$$

and substituting for $f(z)$ from Eq. 6.37 and for $N(r_2,z)$ from Eq. 6.38,

$$\frac{Dp}{RT}2\pi\left[\frac{B}{L_2} - N(r_1)\left(1+\frac{L_1}{L_2}\right) + \Delta A(r_2^2 - r_1^2)N(1-N)\right] = L_1\frac{d}{dz}N(r_1)\ln\frac{r_2}{r_1}$$

Integrating this ordinary differential equation for $N(r_1,z)$ and rearranging the result,

$$\begin{aligned} N(r_1,z) - N(r_1,0) &= (1-\theta)[N(r_2,0) - N(r_1,0) \\ &\quad + \Delta A(r_2^2 - r_1^2)N(1-N)](1-e^{-bz}) \\ N(r_2,z) - N(r_2,0) &= -\theta[N(r_2,0) - N(r_1,0) \\ &\quad + \Delta A(r_2^2 - r_1^2)N(1-N)](1-e^{-bz}) \end{aligned} \tag{6.40}$$

where

$$\begin{aligned} \theta &= \frac{L_1}{L_1 + L_2} \\ b &= \frac{2\pi Dp}{RT}\frac{1}{\theta(1-\theta)\ln(r_2/r_1)}\frac{1}{L_2 + L_1} \end{aligned} \tag{6.41}$$

The separative power when $N(r_2,0)$ and $N(r_1,0)$ are most favorably related[1] is, according to Eq. 11 of Appendix E,

$$\delta U = [L(r_1) + L(r_2)] \frac{\theta(1-\theta)}{2} \frac{(1-e^{-bz})}{(1+e^{-bz})} (\ln \alpha_0)^2 \left(1 - \frac{r_1^2}{r_2^2}\right)^2$$

and the separative power per unit length is

$$\frac{\delta U}{z} = \frac{2\pi Dp}{RT} (\ln \alpha_0)^2 \frac{(1-e^{-bz})}{bz(1+e^{-bz})} \frac{\left(1 - \frac{r_1^2}{r_2^2}\right)^2}{\ln \frac{r_2^2}{r_1^2}}$$

which is a maximum with respect to r_1/r_2 for any value of bz when $r_1/r_2 = 0.534$,

$$\frac{\delta U}{z} = \frac{2\pi Dp}{RT} (\ln \alpha_0)^2 \frac{(1-e^{-bz})}{bz(1+e^{-bz})} 0.40724$$

and a relative maximum with respect to bz when bz = 0, namely,

$$\left(\frac{\delta U}{z}\right)_{max} = 0.40724\pi \frac{Dp}{RT} (\ln \alpha_0)^2 \tag{6.42}$$

If this result is compared with Eq. 6.23, it is seen that this is 81.45 per cent of the theoretical efficiency.

For a single entrance centrifuge, it is necessary to take $N(r_2,0) = N(r_1,0)$. The separative power is then, according to Eq. 8 of Appendix E,

$$\delta U = [L(r_1) + L(r_2)] \frac{\theta(1-\theta)}{2} (1-e^{-bz})^2 (\ln \alpha_0)^2 \left(1 - \frac{r_1^2}{r_2^2}\right)^2$$

and per unit length

$$\frac{\delta U}{z} = \frac{2\pi Dp}{RT} (\ln \alpha_0)^2 \frac{(1-e^{-bz})^2}{bz} \frac{\left(1 - \frac{r_1^2}{r_2^2}\right)^2}{\ln \frac{r_2^2}{r_1^2}}$$

[1] $N(r_2,0) - N(r_1,0) = -\frac{e^{-bz}}{1+e^{-bz}} \Delta A(r_2^2 - r_1^2) N(1-N)$

The maximum with respect to r_1/r_2 is the same as before. The expression[1] $(1 - e^{-bz})^2/bz$ has a true maximum at $bz = 1.25643$ and thus

$$\left(\frac{\delta U}{z}\right)_{max} = (0.40724)^2 \frac{2\pi Dp}{RT} (\ln \alpha_0)^2 \qquad (6.43)$$

which has an efficiency of 66.34 per cent. The condition $bz = 1.25643$ gives a definite relation between the length of the rotor and the flow, namely,

$$L_2 + L_1 = \frac{4\pi Dp}{1.5786RT} \frac{z}{\theta(1 - \theta)} \qquad (6.44)$$

If it is remembered that the separative power is the product of the concentration difference and the transport, the maximum of δU with respect to r_1/r_2 is easily understandable. The concentration difference is the maximum when $r_1/r_2 = 0$, but the transport (through the cylinder of radius zero) is zero. The transport is a maximum when $r_1/r_2 = 1$, but the concentration difference is zero. The maximum of the product thus lies somewhere in between.

A fuller discussion of the significance of the dependence of the results on $1 - e^{-bz}$, the degree of equilibrium attained, is given in Appendix E. The phenomenon is a quite general one.

Notice that b contains the factor $\theta(1 - \theta)$ $(L_1 + L_2)$. Decreasing $\theta(1 - \theta)$ and increasing $L_1 + L_2$ leaves this factor unchanged, but small total flows are desirable and $\theta(1 - \theta)$ should obviously be made as large as possible, which requires $\theta = \frac{1}{2}$ (cf. Eq. 6.44).

Further properties[2] of the concurrent centrifuges may be obtained from the equations given.

[1] Note that the correspondence $e^{-bz} \to r_1^2/r_2^2$ makes this take the form

$$\frac{\left(1 - \frac{r_1^2}{r_2^2}\right)^2}{\ln \frac{r_2^2}{r_1^2}}$$

[2] For example, for any r_1/r_2 the dependence on $z/(L_1 + L_2)$ is the same, and the maximum for the single entrance centrifuge is at the same place, $bz = 1.256$ [with the appropriate value of $\ln (r_2/r_1)$ in b]. But for $z/(L_1 + L_2)$ preassigned (and not bz) the maximum with respect to r_1/r_2 varies.

6. THE COUNTERCURRENT CENTRIFUGE

Countercurrent separating devices in general have the property of multiplying the equilibrium simple-process factor many times in one unit. Because of the nature of the flow, a concentration gradient whose size is limited only by back-diffusion is established in the direction of the flow.

Since a large separation is obtained in one unit, much of the recycling between units which is otherwise necessary to multiply the simple-process factor is avoided. Furthermore, the problem of cascade operation is very considerably simplified since the number of units in series required to effectuate a given fractionation decreases enormously. The cascade becomes broad instead of long and is easily broken down into independent parallel sections. Countercurrent flow is also efficient from a process standpoint because it is possible to maintain maximum separative power through the entire unit.[1] For all these reasons, which have long been appreciated by chemical engineers, countercurrent processes are preferred if at all possible.

The differential expression of Eq. 6.19 becomes for the countercurrent centrifuge

$$p \frac{\partial N}{\partial t} = \frac{Dp}{r} \frac{\partial}{\partial r} \left[r \frac{\partial N}{\partial r} + 2\Delta A r^2 N(1 - N) \right] - pw(r) \frac{\partial N}{\partial z} + Dp \frac{\partial^2 N}{\partial z^2} \qquad (6.45)$$

The solution of Eq. 6.45 may be found by following the standard methods for solution of partial differential equations. This method was utilized in June and July of 1940. However, Eq. 6.45 is particularly simple since the gradients of the concentration in all directions are small. Thus the equation for a countercurrent centrifugal column can also be solved by the method used by Furry, Jones, and Onsager[2] for the thermal-diffusion column. This possibility was first pointed out, and partially worked out, by Bramley.[3] A similar procedure was employed by Martin and Kuhn.[4]

The results of the two methods are of course identical. The second method will be demonstrated here because it is easier and because it

[1]In contrast to concurrent devices, in which $\partial N/\partial r$ changes with z, so that if the separative power is an optimum in one place it is necessarily less in every other place.

[2] Phys. Rev., 55: 1083 (1939).

[3] Science, 92: 427 (1940).

[4] Z. physik. Chem., A189: 219 (1940).

permits the case in which the mole fraction of the product is large to be included without added complications.

The method depends essentially on the fact that $\partial N/\partial r$ is of the order of $N(1-N)2\Delta Ar$, which is a small quantity. Thus the variation of N with respect to r, as compared to the variation of r, p, or pw, may be neglected. For example

$$\int \frac{\partial N}{\partial z}\, pw\, r\, dr \sim \frac{\partial N}{\partial z} \int pw\, r\, dr + \text{terms of the order of } \Delta A^2$$

Consequently Eq. 6.45 may be integrated with respect to r. The first integration gives directly, dropping $\partial^2 N/\partial z^2$ and $\partial N/\partial t$, which are of second order,

$$\frac{\partial N}{\partial r} = -2\Delta A r\, N(1-N) + \frac{1}{r}\frac{\partial N}{\partial z}\int_0^r \frac{pw}{Dp}\, r\, dr \qquad (6.46)$$

The integration constant (a function of z and t) was evaluated by remembering that $r(\partial N/\partial r) \to 0$ as $r \to 0$.

The second integration with respect to r requires a little more artifice in order to get a form in which the integration constant can be evaluated. The boundary conditions are

$$2\pi \int_0^{r_2} \frac{pw}{RT}\, r\, dr = P \qquad (6.47)$$

$$2\pi \frac{\partial}{\partial z}\int_0^{r_2}\left(\frac{pNw}{RT} - \frac{Dp}{RT}\frac{\partial N}{\partial z}\right) r\, dr = -2\pi \int_0^{r_2} \frac{p}{RT}\frac{\partial N}{\partial t}\, r\, dr \qquad (6.48)$$

Equation 6.48 may be transformed by integrating by parts the first term on the left-hand side,

$$2\pi \frac{\partial}{\partial z}\left(N(r_2,z)\int_0^{r_2}\frac{pw}{RT}\, r\, dr - \int_0^{r_2}\frac{\partial N}{\partial r}\, dr \int_0^r \frac{pw}{RT}\, r\, dr\right)$$
$$= 2\pi \frac{\partial}{\partial z}\int_0^{r_2}\frac{Dp}{RT}\frac{\partial N}{\partial z}\, r\, dr - 2\pi\int_0^{r_2}\frac{p}{RT}\frac{\partial N}{\partial t}\, r\, dr \qquad (6.49)$$

Equation 6.49, although it does not give the integration constant that would permit the integration of $\partial N/\partial r$ itself, tells how to integrate

$\frac{\partial N}{\partial r}\int_0^r pw\, r\, dr$, which is just as useful for present purposes. Substituting Eq. 6.46 in Eq. 6.49, it is found that

$$c_6 \frac{\partial N}{\partial t} = c_5 \frac{\partial^2 N}{\partial z^2} - \frac{\partial}{\partial z}\left[PN + c_1 N(1 - N)\right] \tag{6.50}$$

where

$$c_1 = \frac{2\pi}{RT}\frac{(M_2 - M_1)\omega^2}{RT}\int_0^{r_2} r\, dr \int_0^r pw\, r\, dr$$

$$c_2 = \frac{2\pi}{RT} Dp \frac{r_2^2}{2}$$

$$c_3 = \frac{2\pi}{RT}\frac{1}{Dp}\int_0^{r_2}\frac{dr}{r}\left(\int_0^r pw\, r\, dr\right)^2 \tag{6.51}$$

$$c_5 = c_2 + c_3$$

$$c_6 = \frac{2\pi}{RT}\int_0^{r_2} p\, r\, dr = \text{holdup per unit length (moles/cm)}$$

Equation 6.50 has the form of Eq. 2.1, the typical equation for isotope separation in a square cascade or column. The properties of such columns, individually and in cascades, for fixed values of c_1, c_5, and c_6 were studied in Chaps. 2, 3, and 4. The following discussion covers the investigation of how these coefficients vary as a result of changes in operating conditions. The material in the earlier chapters will be referred to constantly.

In order to examine properties of the countercurrent centrifuge, attention will be confined for the moment to a single centrifuge. For convenience it will also be assumed that $N \ll 1$, although this is by no means essential to the argument. Then the steady-state solution of Eq. 6.50 is (Eq. 2.14)

$$\frac{N(Z)}{N(0)} = \frac{1 + \psi \quad \exp\left[2\epsilon Z(1 + \psi)\right]}{1 + \psi \quad \exp\left[2\epsilon Z(1 + \psi)\right]} = \frac{(1 + \psi)e^{2\epsilon S(1+\psi)}}{1 + \psi e^{2\epsilon S(1+\psi)}}$$

where

$$\psi = \frac{P}{c_1} = \frac{PRT}{2\pi (M_2 - M_1) \dfrac{\omega^2}{RT} \displaystyle\int_0^{r_2} r\,dr \int_0^{r} pw\,r\,dr}$$

$$2\epsilon = \frac{c_1}{c_5} = \frac{(M_2 - M_1) \dfrac{\omega^2}{RT} \displaystyle\int_0^{r_2} r\,dr \int_0^{r} pw\,r\,dr}{Dp \dfrac{r_2^2}{2} + \dfrac{1}{Dp} \displaystyle\int_0^{r_2} \frac{dr}{r} \left(\int_0^{r} pw\,r\,dr \right)^2} \tag{6.52}$$

The quantities c_1 and c_5 depend on the function pw, which measures the up-and-down flow, in two ways: They depend on the flow pattern, or the locations and relative size of the streams, and on the absolute magnitude of the flow. The pattern is built into the rotor by the position of orifices in the end caps and position (or absence) of baffles. But the magnitude of the flow is an operating variable and can be adjusted from outside. This independence of flow rate and internal construction in a properly conceived centrifuge gives it more flexibility than the otherwise analogous thermal-diffusion column.

Therefore a new quantity L is introduced.

$$L = \frac{2\pi}{RT} \int_0^{r_2} |pw|\,r\,dr \tag{6.53}$$

which is the flow up the column plus the flow down the column. Then

$$\int_0^{r} \frac{pw\,r\,dr}{L}$$

does not depend on L, and Eq. 6.52 becomes

$$\psi = \frac{P}{a_1 L}$$

$$2\epsilon = \frac{a_1 L}{c_2 + a_3 L^2} \tag{6.54}$$

where $a_1 = c_1/L$ and $a_3 = c_3/L^2$ do not vary with L.

$N(Z)/N(0)$ is now a function of L and ψ. For any value of L, $N(Z)/N(0)$ is greatest when the rate of production is zero. Then

$$\frac{N(Z)}{N(0)} = \exp\left(\frac{a_1 L}{c_2 + a_3 L^2}\, Z \right) \tag{6.55}$$

and the maximum possible value of N(Z)/N(0) for a given length of centrifuge and for a given type of countercurrent flow is

$$\frac{N(Z)}{N(0)} = \exp 2\epsilon_0 Z \tag{6.56}$$

where $2\epsilon_0$ is the maximum of $\frac{a_1 L}{c_2 + a_3 L^2}$ with respect to L and is

$$2\epsilon_0 = \frac{a_1}{2\sqrt{c_2 a_3}} = \frac{(M_2 - M_1)\dfrac{\omega^2}{RT}\displaystyle\int_0^{r_2} r\,dr \int_0^r pw\,r\,dr}{r_2\sqrt{2\displaystyle\int_0^{r_2}\frac{dr}{r}\left(\int_0^r pw\,r\,dr\right)^2}} \tag{6.57}$$

The corresponding value of L is

$$L_0 = \sqrt{\frac{c_2}{a_3}} = \frac{2\pi r_2\left(\dfrac{Dp}{RT}\right)\displaystyle\int_0^{r_2}|pw|\,r\,dr}{\sqrt{2\displaystyle\int_0^{r_2}\frac{dr}{r}\left(\int_0^r pw\,r\,dr\right)^2}} \tag{6.58}$$

The existence of a maximum in ϵ is physically due to the phenomenon of back-diffusion. The concentration gradient is proportional to ϵ. When L, the total flow in the centrifuge, is large and the concentration gradient is small, back-diffusion is negligible ($c_2 \ll a_3 L^2$) and ϵ varies inversely as L. But as soon as L is small and ϵ and the concentration gradient are large, back-diffusion plays an important role, finally limiting the attainable fractionation.

It will be found convenient to measure all flows in terms of L_0. Therefore set

$$m = \frac{L}{L_0} \tag{6.59}$$

Substituting in ϵ,

$$\frac{N(Z)}{N(0)} = \frac{(1+\psi)\exp\left[2\epsilon_0 Z\,\dfrac{2m}{1+m^2}(1+\psi)\right]}{1+\psi\exp\left[2\epsilon_0 Z\,\dfrac{2m}{1+m^2}(1+\psi)\right]} \tag{6.60}$$

In terms of the dimensionless parameters m and ψ, P is given as

$$P = m\psi P_0 \tag{6.61}$$

where

$$P_0 = 4\pi\epsilon_0 r_2^2 \frac{Dp}{RT} \qquad (6.62)$$

Equations 6.57 to 6.62 constitute a complete description of the behavior of the countercurrent centrifuge. When the type of flow pw is specified, ϵ_0, L_0, and P_0 are determined. L and P are the conditions imposed and are measured on flowmeters by the experimenter. By Eqs. 6.59 and 6.61, m and ψ are therefore known. And finally, substituting ϵ_0, m, and ψ in Eq. 6.60, the fractionation of the column is predictable.

The maximum separation factor occurs at $L = L_0$, but this is clearly not the most efficient circulation rate from the standpoint of separative power, since the parasitic phenomenon of back-diffusion is very important at this flow. The relative weight of back-diffusion is reduced by increasing L (thus decreasing the concentration gradient). These qualitative considerations are confirmed by examining δU. From Eq. 2.33,[1]

$$\delta U = \frac{c_1^2}{4c_5} Z = \frac{a_1^2 L^2 Z}{4(c_2 + a_3 L^2)} = \frac{a_1^2}{4a_3} Z\left(\frac{m^2}{1 + m^2}\right) = \frac{c_1^2}{4c_3} Z\left(\frac{m^2}{1 + m^2}\right) \qquad (6.63)$$

whose largest value is reached asymptotically as $m \to \infty$.

It should be noticed that the limiting value of δU is approached very rapidly: When m = 3 the separative power is 90 per cent, and when m = 5 it is 96 per cent of $c_1^2 Z/4c_3$. At these values of m the concentration gradient is 60 and 38 per cent, respectively, of the maximum.[2] Thus it is not necessary to discard the advantages of large separation in one unit, or to increase the flows inordinately, to get adequate separation efficiency.

REFERENCES

Bramley, A., The Centrifugal Separator, Oct. 7, 1940.

Cohen, K., Report A-54, October 1940; Report A-52, Jan. 21, 1941; Report A-50, Jan. 30, 1941.

[1] Consistent with the earlier sections of this chapter, the symbol δU is being used for the total separative power, and not the separative power per unit length as in Chap. 2.

Cohen, K., memorandum to H. C. Urey, May 14, 1940.
Cohen, K., and I. Kaplan, Report A-101, Jan. 28, 1942.
Cohen, K., and C. Skarstrom, Columbia Ser. No. 4R-X138, Feb. 6, 1941.
Dirac, P. A. M., British MS, 1941.
Kaplan, I., and K. Cohen, Report A-195, June 17, 1942.
Murphree, E. V., Separation of Gases by Diffusional Methods, February 1942.

Chapter 7

OTHER SEPARATION METHODS

1. TWO-PHASE SEPARATIONS

The classic case of a two-phase separation is fractional distillation, one of the most widely used unit operations of chemical industry. The earliest use of this process in isotope separation was the fractionation of Ne^{20} from liquid neon by Keesom. The method was later used to separate deuterium from liquid hydrogen and in the fractionation of H_2O to give O^{18} (Urey et al.).

The scope of this process was much enlarged by Urey's invention of the chemical-exchange method. Here the driving force of the separation is the chemical equilibrium constant of a reaction such as

$$HC^{13}N(gas) + C^{12}N^-(aq) \rightleftharpoons HC^{12}N(gas) + C^{13}N^-(aq)$$

or

$$N^{15}H_3(gas) + N^{14}H_4^+(aq) \rightleftharpoons N^{14}H_3(gas) + N^{15}H_4^+(aq)$$

Such a reaction can be carried out entirely analogously to ordinary fractional distillation (refer to Fig. 7.1). In the example shown, HCN gas is scrubbed countercurrently with a water solution of NaCN in a packed tower, just as vapor and liquid are contacted in fractional distillation. The condenser of the fractionating tower is replaced by NaOH, which converts the gas phase to the liquid phase.

Although the theory of such processes was fairly well understood before the war, an account of it is included here by reason of its probable great future importance. Also, the generality of development of cascade theory now allows the theory to be given in a clearer and more concise form. The theory has some points of resemblance with the ordinary theory of fractional distillation. The two chief differences are (1) the closed form of the results, which is possible because isotopic solutions are perfect solutions and because the simple-process factor is nearly one; and (2) the inclusion of the equilibrium-time phenomenon, which is generally omitted although it is more often significant than is commonly suspected.

It is clear that the two phases need not be liquid and gas; two liquids, or a liquid and a solid, would be equally suitable. The phase

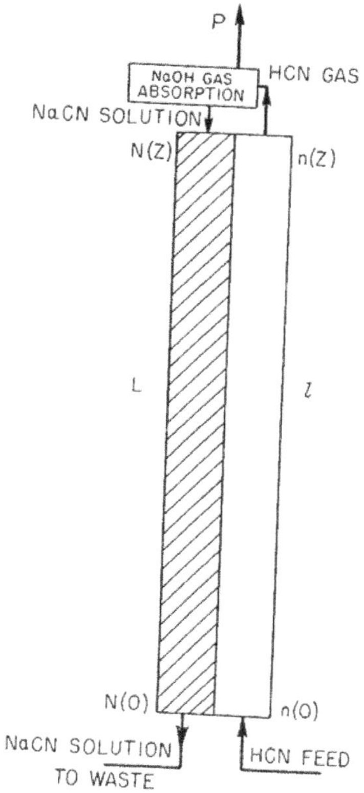

Fig. 7.1—A chemical-exchange tower.

in which enrichment of desired isotope takes place will be denoted by small letters and the other phase by capitals. Thus n, l, h, and d are, respectively, the mole fraction, flow in moles per second, holdup in moles per centimeter, and diffusion coefficient in one phase; and N, L, H, and D are the corresponding quantities for the other phase. At equilibrium[1]

[1]A complete discussion of the theory of the equilibrium constant may be found in J. Bigeleisen and M. G. Mayer, J. Chem. Phys., 15: 261 (1947).

$$\left(\frac{n}{1-n}\right)\left(\frac{1-N}{N}\right) = \alpha > 1 \tag{7.1}$$

Countercurrent contacting (Fig. 7.1) will be considered first. The direction of flow of the enriched phase will be taken as positive, and length along the column will be measured in this direction. Then if T is the rate of transfer of desired isotope to the enriched phase across the interface per unit length of column,

$$\begin{aligned} h\frac{\partial n}{\partial t} + l\frac{\partial n}{\partial z} &= +T \\ H\frac{\partial n}{\partial t} - L\frac{\partial N}{\partial z} &= -T \end{aligned} \tag{7.2}$$

The rate of transfer T depends on two processes: the rate of attainment of the chemical equilibrium and the rate of diffusion through fluid boundary layers. If the equilibrium process is rate determining, then T has the form of the rate constant for a bimolecular reversible reaction

$$T = -k'cC\,[n(1-N) - \alpha N(1-n)] \tag{7.3}$$

where c and C are the concentrations in moles per liter of the process material in two phases.

Usually the slow process is the diffusion through the boundary layers. Referring to Fig. 7.2 and denoting the mole fractions at the interface by n_i and N_i, and the area of the interface per unit length of column by σ, then

$$\frac{T}{\sigma} = -DC\frac{\partial N_i}{\partial x} = -dc\frac{\partial n_i}{\partial x} \tag{7.4}$$

The differential coefficients may be replaced by $(N_i - N)/B$ and $(n - n_i)/b$, respectively, when B and b are the film thicknesses. Thus

$$\frac{T}{\sigma} = -\frac{DC}{B}(N_i - N) = -\frac{dc}{b}(n - n_i) \tag{7.5}$$

At the same time n_i and N_i are related by Eq. 7.1, so that

$$n_i - N_i = (\alpha - 1)N_i(1 - n_i) \tag{7.6}$$

Eliminating n_i and N_i between Eqs. 7.5 and 7.6,

$$T = -k\,[(n - N) - (\alpha - 1)N_i(1 - n_i)] \tag{7.7}$$

where

$$\frac{1}{k} = \frac{B}{DC\sigma} + \frac{b}{dc\sigma} \tag{7.8}$$

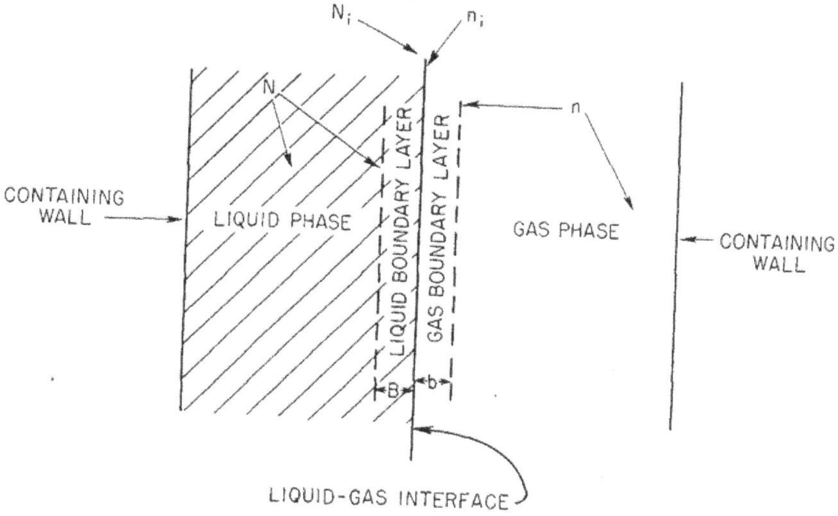

Fig. 7.2—Relations at a gas-liquid interface.

a relation that is quite familiar to chemical engineers. Equation 7.7 can be simplified by the assumption that the interfacial concentrations do not differ appreciably from the concentrations in the body of the fluid, giving

$$\begin{aligned} T &= -k\,[(n - N) - (\alpha - 1)N(1 - n)] \\ &= -k\,[n(1 - N) - \alpha N(1 - n)] \end{aligned} \tag{7.9}$$

which is the same as Eq. 7.3 but with a different meaning for k. It should be noted that in deriving Eq. 7.9 it has not been assumed that $(\alpha - 1)$ is small.

Inserting Eq. 7.9 in Eq. 7.2,

$$h\frac{\partial n}{\partial t} + l\frac{\partial n}{\partial z} = -k\left[(n - N) - (\alpha - 1)N(1 - n)\right]$$

$$H\frac{\partial N}{\partial t} - L\frac{\partial N}{\partial z} = k\left[(n - N) - (\alpha - 1)N(1 - n)\right] \qquad (7.10)$$

These two equations may now be combined, using the assumption that $(\alpha - 1)$ is very small (for the case of large values of α, see Sec. 3). Noting that the operator $\partial/\partial t$ is proportional to $(\alpha - 1)^2$ and that $\partial/\partial z$ is of the order $(\alpha - 1)$,

$$(H + h)\frac{\partial n}{\partial t} = -\frac{\partial}{\partial z}\left[\frac{Ll}{k}\frac{\partial n}{\partial z} - L(\alpha - 1)n(1 - n) + Pn\right] \qquad (7.11)$$

which is the form of the fundamental equation of isotope separation (Eq. 2.1) with the correspondences

$$c_6 \rightarrow H + h$$
$$c_5 \rightarrow -Ll/k \qquad (7.12)$$
$$c_1 \rightarrow -L(\alpha - 1)$$

The theory of square cascades developed in Chap. 2 can thus be applied directly.

The exact solution of Eq. 7.10 in the steady state for values of $(\alpha - 1)$ that are not small is also of some interest. Eliminating N by the relation $ln - LN = Pn_p$, defining $\psi = P/(\alpha - 1)l$, $\epsilon = k(\alpha - 1)/2L$, the following is obtained:

$$Z = \frac{1}{\epsilon\,\Delta(\gamma)}\tanh^{-1}\left\{\frac{(n_z - n_0)\,\Delta(\psi)}{(n_z + n_0)\left[1 + \psi + (\alpha - 1)\psi n_p\right] - 2n_z n_0 - 2n_p\alpha\psi}\right\} \qquad (7.13)$$

where $\Delta(\psi) = \left\{[1 + \psi + (\alpha - 1)\psi n_p]^2 - 4\alpha\psi n_p\right\}^{1/2}$ and n_0, n_z, and n_p are the mole fractions at the bottom of the column, the top of the column, and the top of the cascade, respectively. Equation 7.13 is directly analogous to Eq. 2.5, to which it reduces for small values of $(\alpha - 1)$.

The equations for concurrent contacting are easily obtained from Eq. 7.10 by changing L to $-L$. This gives

$$H\frac{\partial N}{\partial t} + L\frac{\partial N}{\partial z} = k\left[(n - N) - (\alpha - 1)N(1 - n)\right]$$

$$h\frac{\partial n}{\partial t} + l\frac{\partial n}{\partial z} = -k\left[(n - N) - (\alpha - 1)N(1 - n)\right] \qquad (7.14)$$

The solution of these equations in the steady state ($\partial N/\partial t = \partial n/\partial t = 0$) for ($\alpha - 1$) small has been treated at length in Appendix E, as well as the associated cascading problem.

2. THERMAL DIFFUSION

The history and theory of the thermal-diffusion process has been the subject of a number of recent papers which have been widely read.[1] The treatment here, therefore, will be confined to a rapid review of the theory to show how it fits in with other separation methods in the general scheme.

The equation for thermal diffusion corresponding to Eq. 6.45 that was derived for a centrifuge is easily arrived at by replacing

$$\left[\frac{\partial N}{\partial r} + \frac{(M_2 - M_1)\omega^2 r}{RT} N(1 - N)\right]$$

by the corresponding expression for thermal diffusion[2]

$$\left[\frac{\partial N}{\partial x} - \frac{\alpha}{T}\frac{\partial T}{\partial x} N(1 - N)\right]$$

where α is known as the thermal-diffusion coefficient.

Then, in place of Eq. 6.45,

$$\rho\frac{\partial N}{\partial t} = \frac{\partial}{\partial x} D\rho\left[\frac{\partial N}{\partial x} - \frac{\alpha}{T}\frac{\partial T}{\partial x} N(1 - N)\right] - \rho\omega\frac{\partial N}{\partial z} + D\rho\frac{\partial^2 N}{\partial z^2} \qquad (7.15)$$

where x is the distance from one wall in a thin annular space between two concentric cylinders. It must be remembered that because of the severe temperature drop across the annulus ρD varies with x.

Proceeding now precisely as in Sec. 6 of Chap. 6, it is found that Eq. 7.15 resolves to

$$c_6\frac{\partial N}{\partial t} = c_5\frac{\partial^2 N}{\partial z^2} - \frac{\partial}{\partial z}\left[PN + c_1 N(1 - N)\right] \qquad (7.16)$$

[1] L. Waldmann, Z. Physik, 114: 53 (1939); H. Furry, R. C. Jones, and L. Onsager, Phys. Rev., 55: 1083 (1939); P. Debye, Ann. Physik, 36: 284 (1939); K. Wirtz, Ann. Physik, 36: 295 (1939); A Bramley, Science, 92: 477 (1940); J. Bardeen, Phys. Rev., 57: 35 (1940), 58: 94 (1940).

[2] See S. Chapman and T. G. Cowling, "Mathematical Theory of Non-Uniform Gases," p. 244, Cambridge University Press, London, 1939.

where now

$$c_1 = -2\pi r \int_0^a \frac{\alpha}{T}\frac{\partial T}{\partial x}\int_0^x \rho\omega\, dx$$

$$c_2 = 2\pi r \int_0^a D\rho\, dx$$

$$c_3 = 2\pi r \int_0^a \frac{dx}{\rho D}\left(\int_0^x \rho\omega\, dx\right)^2 \qquad (7.17)$$

$$P = 2\pi r \int_0^a \rho\omega\, dx = \text{grams/sec if } \rho \text{ is in grams/cu cm}$$

$$c_5 = c_2 + c_3$$

$$c_6 = 2\pi r \int_0^a \rho\, dx$$

In Eq. 7.17, ρ is the density, r is the radius of the annulus, and a is the spacing across the annular gap. The relations of Eqs. 7.16 and 7.17 hold for any fluid, liquid as well as gas.

Values of c_1 and c_5 depend on the spacing, the temperature drop across the gap, and the intrinsic properties of the fluid.

The further theory of the thermal-diffusion column closely resembles that of the centrifuge, except that in the present instance the mass flow is caused by convection and is dependent on the temperature drop and the spacing. Corresponding to ϕ_0, the flow that gives the maximum fractionating factor in a centrifuge, there is a characteristic width a_0 for which the fractionation in a thermal-diffusion column is a maximum. However the maximum efficiency from the standpoint of separative work per unit of heat loss is at a spacing other than a_0 which corresponds to the maximum separative efficiency in a centrifuge for flows larger than ϕ_0.

From the standpoint of flow geometry, it is easy to show, by the method of Chap. 6, Sec. 6, that the model described has an absolute efficiency of about 71 per cent and that the most efficient flow pattern corresponds to flow immediately adjacent to the walls of the annulus, but with no flow in the middle.

3. CONCENTRATION OF DEUTERIUM

A possible basis for the production of deuterium is the exchange reaction which requires a catalyst. It was first proposed by Urey in late 1940 to use this process by countercurrent washing of water and hydrogen in a tower packed with catalyst. At the time of writing, water must be converted into hydrogen at the product end of a cascade using the process of Eq. 7.17.

The steady-state equations corresponding to Eq. 1 of Appendix E are, from Eq. 6.14,[1]

$$
\begin{aligned}
L \frac{\partial N}{\partial z} &= k\,[(n - N) - (\alpha - 1)N(1 - n)] \\
l \frac{\partial n}{\partial z} &= -k\,[(n - N) - (\alpha - 1)N(1 - n)]
\end{aligned}
\tag{7.19}
$$

Appreciable simplification occurs when $N \ll 1$; since this is the major part of the D_2 problem (initial concentration = 1/7,000) this case will be considered. Then

$$
\begin{aligned}
L \frac{\partial N}{\partial z} &= k(n - \alpha N) \\
l \frac{\partial n}{\partial z} &= -k(n - \alpha N)
\end{aligned}
\tag{7.20}
$$

The solutions of Eq. 7.20 are

$$
\begin{aligned}
n(Z) &= -\frac{[\alpha N(0) - n(0)]L}{L + \alpha l} \exp\left[-\frac{k(L + \alpha l)Z}{lL}\right] + \frac{\alpha[n(0)l + N(0)L]}{L + \alpha l} \\
N(Z) &= \frac{[\alpha N(0) - n(0)]l}{L + \alpha l} \exp\left[-\frac{k(L + \alpha l)Z}{lL}\right] + \frac{n(0)l + N(0)L}{L + \alpha l}
\end{aligned}
\tag{7.21}
$$

The cascade is shown in Fig. 7.3. Then for the (s + 1)th stage, the following identifications can be made:

$$
N(0) \rightarrow N_s \qquad N(Z) \rightarrow N_{s+1} \qquad n(0) \rightarrow n_s \qquad n(Z) \rightarrow n_{s-1}
$$

Equation 7.21 becomes

$$
\begin{aligned}
n_{s-1} &= \frac{N_s \alpha\beta L}{L + \alpha l} + n_s\left(1 - \frac{L\beta}{L + \alpha l}\right) \\
N_{s+1} &= N_s\left(1 - \frac{\alpha\beta l}{L + \alpha l}\right) + \frac{n_s l\beta}{L + \alpha l}
\end{aligned}
\tag{7.22}
$$

[1]Since $(\alpha - 1)$ is large, we may no longer set $N(1 - n) = N(1 - N)$.

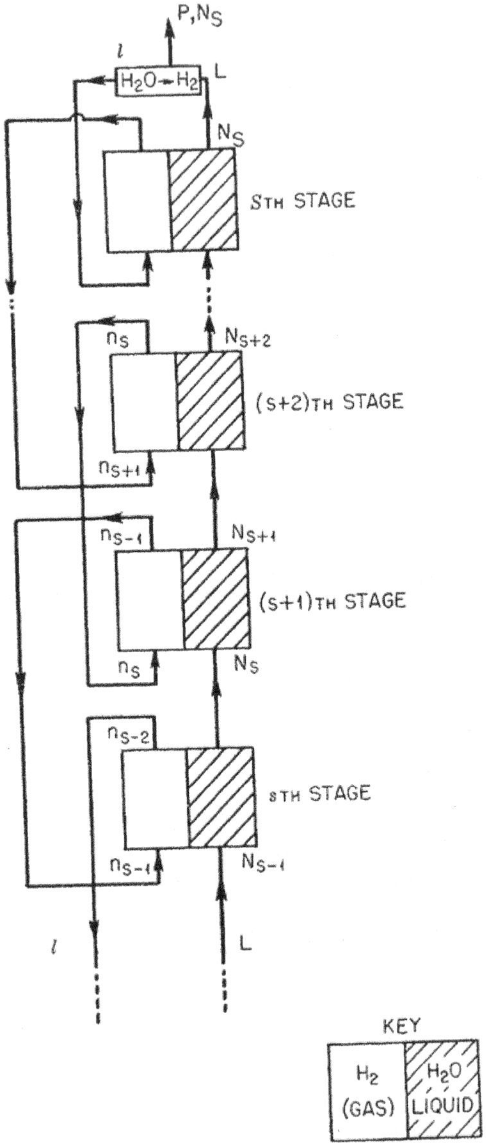

Fig. 7.3—Schematic representation of a heavy-water cascade for the catalyst-exchange method.

where

$$\beta \equiv 1 - \exp\left[-\frac{k(L + \alpha l)Z}{lL}\right] = \text{fraction of equilibrium attained} \tag{7.23}$$

Now Eq. 7.22 can be combined with the conservation equation

$$LN_S - l\,n_{S-1} = PN_S \tag{7.24}$$

to give three equations in the four quantities N_{S+1}, N_S, n_S, and n_{S-1} which can be solved in terms of any one of them. For example

$$N_{S+1} = N_S \frac{L + \alpha l(1 - \beta)}{L(1 - \beta) + \alpha l} - \frac{\beta PN_S}{L(1 - \beta) + \alpha l} \tag{7.25}$$

The difference equation (Eq. 7.25) is easily solved to give

$$\frac{N_S}{N_0} = A^S - B\,\frac{A^S - 1}{A - 1} \tag{7.26}$$

where

$$A \equiv \frac{L + \alpha l(1 - \beta)}{L(1 - \beta) + \alpha l}$$

$$B \equiv \frac{\beta PN_S}{L(1 - \beta) + \alpha l} \tag{7.27}$$

and N_0 is the concentration of HDO in the feed. The number of sections in the cascade, as a function of the initial and final concentration, is accordingly

$$S = \frac{\ln\left(\frac{N_S}{N_0} - \frac{B}{A - 1}\right) - \ln\left(1 - \frac{B}{A - 1}\right)}{\ln A} \tag{7.28}$$

Since $L - l = P$, A and B (and hence S) are functions of P/L, the percentage withdrawal.[1] The volume of catalyst required can be obtained by writing

$$kZ = k'V$$

where V is the volume of catalyst per stage and k′ is the transfer coefficient per unit of catalyst. Then from Eq. 7.23,

$$V = -\frac{lL}{(L + \alpha l)k'} \ln (1 - \beta) \tag{7.29}$$

and the total volume is VS, a function of β and P. The total volume of catalyst has its smallest value when β, the fractional attainment of equilibrium, is zero. Without proof, this limiting volume is taken to be the same as that for the operation of a countercurrent column and corresponds to an infinite number of stages. The minimum number of stages occurs for $\beta = 1$ and of course requires an infinite amount of catalyst.

It can also be shown that there is a maximum in the function

$$\frac{P}{VS} = \frac{\text{production}}{\text{volume of catalyst}}$$

which gives the economic operating conditions when the catalyst volume is limiting. However, it is usually expedient to run at higher values of P and get more complete recovery of heavy water from feed.

REFERENCES

Cohen, K., Report A-118, May 1941; Report A-530, Feb. 4, 1943; Report A-531, Feb. 5, 1943.

Cohen, K., Columbia Ser. No. 4M-X148, June 25, 1941; Columbia Ser. No. 4R-X133, August 1941.

Cohen, K., Columbia Ser. No. 4L-35, Mar. 17, 1943 (letter to P. C. Keith).

Cohen, K., Packed Fractionating Columns, J. Chem. Phys., 8: 588 (1940).

Cohen, K., and I. Kaplan, Columbia Ser. No. 4R-6, Nov. 5, 1942.

Hutchison, C. A., Jr., "Chemical Separation of the Uranium Isotopes," Division III, Volume 3, National Nuclear Energy Series.

Kaplan, I., and K. Cohen, Report A-335, Oct. 19, 1942; Report A-1276, May 1, 1944.

Thompson, W. I., and K. Cohen, Ser. No. 4R-104, Sept. 3, 1943.

APPENDIXES

Appendix A — Roots of a Transcendental Equation

Appendix B — Equilibrium Time of Square Cascades for $N \sim 1$

Appendix C — The Holdup Function

Appendix D — Rayleigh Distillation

Appendix E — Properties of Concurrent Two-phase Elements

Appendix A

ROOTS OF A TRANSCENDENTAL EQUATION

The features of the solution of the characteristic relations of Eq. 3.20 or Eq. 3.22 for the roots p_k will be made clear by completely discussing the simplest special case, namely $K' = \infty$, $K = 0$, and $P = 0$. In this case it is necessary to solve for all the roots p_k of the equation

$$\tan \gamma S = \frac{\gamma S}{\epsilon S} \tag{1}$$

where

$$\gamma = i\beta = i\sqrt{\epsilon^2 + \lambda p} \tag{2}$$

or

$$p = -\frac{1}{\lambda S^2}\left[(\epsilon S)^2 + (\gamma S)^2\right] \tag{3}$$

If both sides of Eq. 1 are plotted as functions of γS (Fig. 1), the solutions of the equation are the intersections of the curves $y = \tan \gamma S$ and $y = \gamma S/\epsilon S$. If ϵS is positive and less than 1, there is a root $\gamma S = (\gamma S)_1$ in the first branch of the tangent function, as well as other roots which approximate to

$$\frac{(2K-1)\pi}{2} \quad (K = 2,3,4 \ldots)$$

in all succeeding branches. Since p depends on $(\gamma S)^2$, the symmetrical negative roots may be neglected. The smallest root is in the first branch and gives p_1; the other roots in the later branches give p_2, p_3 . . . .

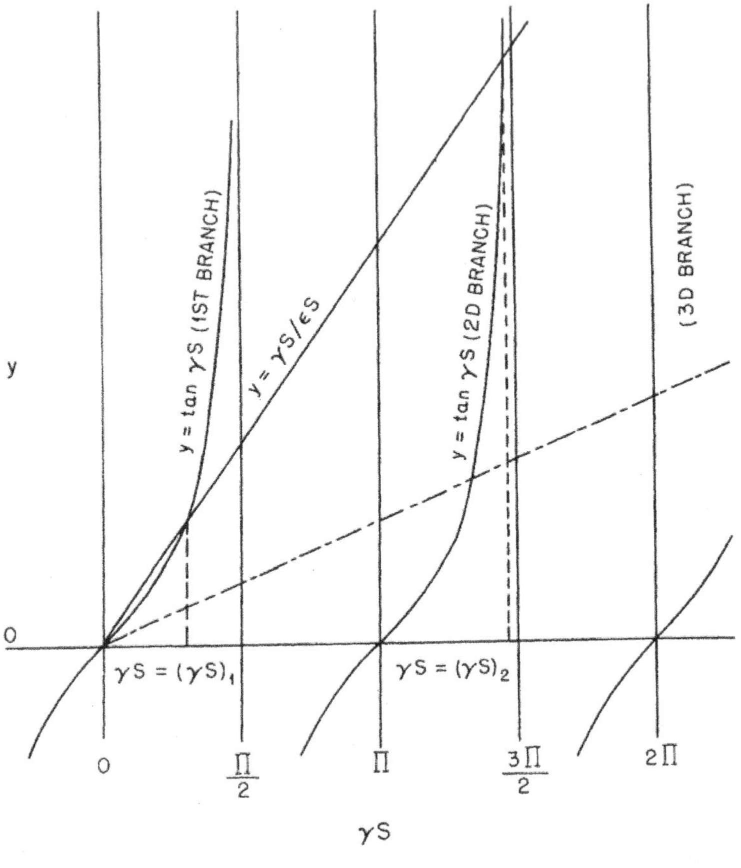

Fig. 1—Graph of both sides of the equation tan $\gamma S = \gamma S/\epsilon S$ as functions of γS.

If ϵS is positive and greater than 1, the root in the first branch disappears (dash-dotted line, Fig. 1). In this case there is a pure imaginary root, $\gamma_1 = i\beta_1$. Substituting β in Eq. 1,

$$\tanh \beta S = \frac{\beta S}{\epsilon S} \tag{4}$$

and

$$p = -\frac{1}{\lambda S^2} \left[(\epsilon S)^2 - (\beta S)^2\right] \tag{5}$$

Plotting both sides of Eq. 4 as functions of βS (Fig. 2), a root of $(\beta S)_1$ is observed. Since $\tanh (\beta S)_1$ is always less than 1, so also is $(\beta S)_1/\epsilon S$, and consequently p_1 is negative.

For negative ϵS, there is no root of Eq. 4. All the roots γS are real and satisfy Eq. 1. Referring to Fig. 1, there is a root $(\gamma S)_1$ between $\pi/2$ and π in the first branch of the tangent, and there are later roots which approximate to $(2K - 1)\pi/2$ $(K = 2,3,4 \ldots)$ in all succeeding branches. Again, p is negative.

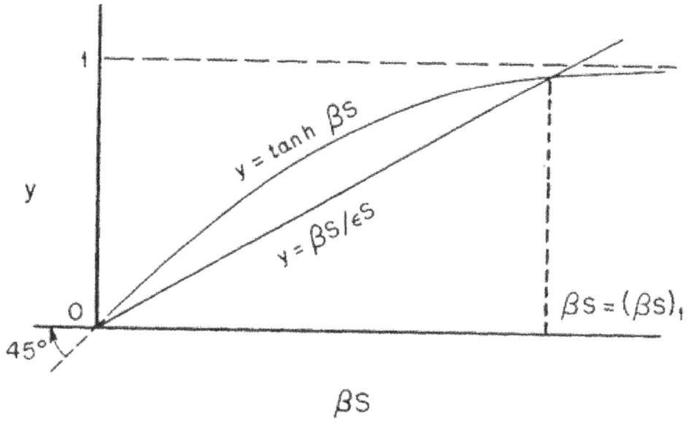

Fig. 2—Graph of both sides of the equation $\tanh \beta S = \beta S/\epsilon S$ as functions of βS.

Approximate expressions may be obtained for the roots $(\gamma S)_1$ when ϵS (of either sign) is small or large, or when S is positive and nearly 1.

<u>ϵS Large and Positive</u>. The value of $\tanh (\beta S)_1$ is very nearly 1. Equation 4 is expressed as

$$\tanh \beta S = 1 - 2e^{-2\beta S} + 2e^{-4\beta S} + \ldots = \frac{\beta S}{\epsilon S} \tag{6}$$

Solving by successive approximations,

$$(\beta S)_1 = \epsilon S[1 - 2e^{-2\epsilon S} - (8\epsilon S - 2)e^{-4\epsilon S} + \ldots] \tag{7}$$

Equation 7 to the number of terms given is good to 0.002 for $\epsilon S > 2$.

<u>ϵS Large and Negative</u>. The value of $(\gamma S)_1$ is nearly π. We set $(\gamma S)_1 = \pi + x$. Then Eq. 1 becomes

$$\tan (\pi + x) = x + \frac{x^3}{3} + \frac{2x^5}{15} + \ldots = \frac{\pi + x}{\epsilon S} \tag{8}$$

and solving by successive approximations,

$$(\gamma S)_1 = \pi\left[1 + \frac{1}{(\epsilon S - 1)} - \frac{(\epsilon S)\pi^2}{3(\epsilon S - 1)^4} + \ldots\right] \tag{9}$$

Equation 9 also gives $(\gamma S)_2$ for ϵS very large and positive, and to the number of terms given it is good to 0.06 for $-\epsilon S > 2$.

ϵS Small, Positive or Negative. The value of tan $(\gamma S)_1$ is nearly infinite, and $(\gamma S)_1$ is nearly $\pi/2$. Set $(\gamma S)_1 = (\pi/2) - x$. Then Eq. 1 becomes

$$\cot x = \frac{\frac{\pi}{2} - x}{\epsilon S} = \frac{1}{x} - \frac{x}{3} - \frac{x^3}{45} - \ldots \tag{10}$$

expanding the cotangent. The size of x is very small and the root is accordingly

$$(\gamma S)_1 = \frac{\pi}{2} - \frac{\epsilon S}{\frac{\pi}{2} - \frac{2}{\pi}\left[\epsilon S - \frac{1}{3}(\epsilon S)^2\right]} + \ldots \tag{11}$$

Equation 11 to the number of terms given is good to 0.001 for $|\epsilon S| \leq 0.20$. The expression also gives the succeeding roots $(\gamma S)_K$ by substituting $(2K - 1)\pi/2$ for $\pi/2$. Explicitly,

$$(\gamma S)_K = (2k - 1)\frac{\pi}{2} - \frac{\epsilon S}{(2k - 1)\frac{\pi}{2} - \frac{2}{(2k - 1)\pi}\left[\epsilon S - \frac{1}{3}(\epsilon S)^2\right]} \tag{12}$$

ϵS Nearly 1. Since $(\gamma S)_1$ is small, Eq. 1 may be written as

$$\tan \gamma S = \gamma S + \frac{(\gamma S)^3}{3} + \frac{2(\gamma S)^5}{15} + \frac{17(\gamma S)^7}{315} + \ldots = \frac{\gamma S}{\epsilon S} \tag{13}$$

whose solution is

$$(\gamma S)_1 = \sqrt{3\left(\frac{1}{\epsilon S} - 1\right) - \frac{18}{5}\left(\frac{1}{\epsilon S} - 1\right)^2 + 4.27\left(\frac{1}{\epsilon S} - 1\right)^3} \tag{14}$$

Table 1 — Roots of the Equation $\gamma S = (\epsilon S) \tan \gamma S$

ϵS	γS	$-\epsilon S$	γS
0.00	1.57080	0.00	1.57080
0.05	1.53830	0.05	1.60200
0.10	1.50443	0.10	1.63199
0.15	1.46904	0.15	1.66087
0.20	1.43203	0.20	1.68868
0.25	1.39325	0.25	1.71551
0.30	1.35251	0.30	1.74140
0.35	1.30965	0.35	1.76641
0.40	1.26440	0.40	1.79058
0.45	1.21649	0.45	1.81396
0.50	1.16556	0.50	1.83660
0.55	1.11118	0.55	1.85852
0.60	1.05279	0.60	1.87976
0.65	0.98966	0.65	1.90036
0.70	0.92079	0.70	1.92035
0.75	0.84473	0.75	1.93974
0.80	0.75938	0.80	1.95858
0.85	0.66086	0.85	1.97687
0.90	0.54228	0.90	1.99465
0.95	0.38537	0.95	2.01194
1.00	0.00000	1.00	2.02876
1.05	0.38924i	1.05	2.04512
1.10	0.55323i	1.10	2.06105
1.20	0.79028i	1.20	2.09166
1.30	0.97768i	1.30	2.12071
1.40	1.14036i	1.40	2.14834
1.50	1.28783i	1.50	2.17463
1.60	1.42503i	1.60	2.19968
1.70	1.55477i	1.70	2.22354
1.80	1.67889i	1.80	2.24633
1.90	1.79867i	1.90	2.26811
2.00	1.91501i	2.00	2.28893

Equation 14, derived here for $\epsilon S < 1$, can be shown to hold for $\epsilon S > 1$, giving the expected imaginary values for $(\gamma S)_1$. It holds to 0.001 for $|\epsilon S - 1| \lesssim 0.1$.

Roots of Eq. 1 for $\epsilon S = [-2.0(0.1) - 1.0; -1.0(0.5) + 1.0; 1.0(0.1)2.0; 5D]$ are listed in Table 1.

REFERENCES

Cohen, K., Packed Fractionating Columns and the Concentration of Isotopes, J. Chem. Phys., 8: 588 (1940).

Cohen, K., Report A-530, Feb. 4, 1943.

Appendix B

EQUILIBRIUM TIME OF SQUARE CASCADES FOR N ~ 1

This section outlines the procedure that will solve Eq. 2.1 in the general case when N is not small. The differential equation is, when P = 0,

$$\lambda \frac{\partial N}{\partial t} = \frac{\partial}{\partial s}\left[\frac{\partial N}{\partial s} - 2\epsilon N(1 - N)\right] \tag{1}$$

Consider the simple case (K′ = ∞, K = 0, P = 0)—a cascade operated under total reflux, with no reservoir on top, and the concentration at the bottom maintained constant. The boundary conditions are

$$\begin{aligned} &\text{At } t = 0, && N = N_0 \\ &\text{At } s = 0, && N = N_0 \\ &\text{At } s = S, && \frac{\partial N}{\partial s} = 2\epsilon N(1 - N) \end{aligned} \tag{2}$$

The substitution of R instead of N as dependent variable makes the boundary conditions linear.

$$\begin{aligned} &\text{At } t = 0, && R = R_0 \\ &\text{At } s = 0, && R = R_0 \\ &\text{At } s = S, && \frac{\partial R}{\partial s} = 2\epsilon R \end{aligned} \tag{3}$$

At the same time, Eq. 1 takes the form

$$\lambda \frac{\partial R}{\partial t} = \frac{\partial^2 R}{\partial s^2} - 2\epsilon \frac{\partial R}{\partial s} + \frac{2}{1 + R}\frac{\partial R}{\partial s}\left(2\epsilon R - \frac{\partial R}{\partial s}\right) \tag{4}$$

Now it is known from the theory of partial differential equations[1] that equations of the form

$$\lambda \frac{\partial R}{\partial t} - \frac{\partial^2 R}{\partial s^2} = f(R, \partial R/\partial s, s, t)$$

with boundary conditions of the type in Eq. 3 are solvable by iteration. It remains to choose the most efficient procedure.

The expression in parentheses on the right-hand side of Eq. 4 is, in terms of N,

$$2\epsilon R - \frac{\partial R}{\partial s} = \frac{1}{(1-N)^2}\left[2\epsilon N(1-N) - \frac{\partial N}{\partial s}\right] = \frac{2(\tau - NT)}{L(1-N)^2}$$

according to Eq. 1.77, where $\tau - NT$ is the net transport of desired material. Since $\partial R/\partial s$ is always positive and since the transport for the initial conditions chosen is also positive, the entire expression

$$\frac{2}{1+R}\frac{\partial R}{\partial s}\left(2\epsilon R - \frac{\partial R}{\partial s}\right)$$

is essentially positive. Thus an iteration process in which we begin by overestimating this term will converge to the solution from low estimates of the equilibrium time; whereas if we begin by underestimating the term we will overestimate the equilibrium time.

The expression $\frac{2}{1+R}\frac{\partial R}{\partial s}$ is the difficult term; the most efficient iteration would seem to be performed on this expression alone, getting successive equations of the form

$$\lambda \frac{\partial R^k}{\partial t} = \frac{\partial^2 R^k}{\partial s^2} - 2\epsilon \frac{\partial R^k}{\partial s} + f_k(s,t)\left(2\epsilon R^k - \frac{\partial R^k}{\partial s}\right) \tag{5}$$

to solve, where R^k is the kth approximation to R and

$$f_k(s,t) = \frac{2}{1+R^{k-1}}\frac{\partial R^{k-1}}{\partial s}$$

[1]Maurice Gevrey, "Sur les équations aux dérivées partielles du type parabolique," Gauthier-Villars, Paris, 1913.

We might take for f_0 either zero or $4\epsilon R_0 e^{2\epsilon s}/(1 + R_0 e^{2\epsilon s})$ corresponding to the initial and final values of R and $\partial R/\partial s$ (underestimates and overestimates of f).

However, unless f_k is particularly simple, Eq. 5 also has to be solved by iteration. Thus practically the solution by way of Eq. 5 fails after R^0. Even the choice $f_0 = 4\epsilon R_0 e^{2\epsilon s}/(1 + R_0 e^{2\epsilon s})$ is too cumbersome; first- or second-degree rational algebraic functions of s which lead to special cases of the confluent hypergeometric equation are more convenient.

An iteration method that can be continued indefinitely is obtained by solving equations of the form

$$\lambda \frac{\partial R^k}{\partial t} = \frac{\partial^2 R^k}{\partial s^2} - 2\epsilon \frac{\partial R^k}{\partial s} + g_k(s,t) \tag{6}$$

where

$$g_k(s,t) = \frac{2}{1 + R^{k-1}} \frac{\partial R^{k-1}}{\partial s} \left(2\epsilon R^{k-1} - \frac{\partial R^{k-1}}{\partial s} \right)$$

Here the homogeneous part of the equation has constant coefficients, which permits the particular solutions to be obtained operationally.

Appendix C

THE HOLDUP FUNCTION

The flow in any section of an ideal cascade is equal to $2\Delta U/\epsilon^2$. Over a small section $\Delta U \sim d^2V/dN^2$. Consider, instead of $U = PV$, the expression $H = PhK$, where h is the holdup per unit flow and

$$\frac{d^2K}{dN^2} \equiv N\frac{d^2V}{dN^2} = \frac{1}{N(1-N)^2} \qquad (1)$$

it is clear that $(2/\epsilon^2)\,\Delta H$ for the small section will be equal to the holdup of desired isotope. Now a large cascade section may be considered as composed of small sections, the product of one section being the feed of the next. Likewise ΔH for the large section is the sum of the ΔH's of the small sections. Hence $(2/\epsilon^2)\Delta H$ over any section of an ideal cascade, large or small, represents the desired material holdup in that section.

The integral of Eq. 1 with the constants chosen so that $K(0.50) = dK(0.50)/dN = 0$ is

$$K = N \ln R + (1 - 2N) \qquad (2)$$

This is called the "elementary holdup function." It is the analogue of V' (not V).

To illustrate the use of H, consider the net desired material holdup in the cascade section B of Fig. 4.6.

Evaluating $\Delta U'$ around this section,

$$\Delta U' = P_B\, V'(N_B) + l_B\, V'(n_B) - L_B\, V'(N_A)$$

Expanding

$$V'(n_B) = V'(N_A) + (n_B - N_A)\frac{d}{dN} V'(N_A)$$

gives the expression

$$\Delta U' = P_B\, V'(N_B) + (l_B - L_B)\, V'(N_A) + l_B(n_B - N_A)\frac{d}{dN} V'(N_A)$$

From Eqs. 4.39 and 4.40, this reduces to

$$\Delta U' = P_B\, V'(N_B) - P_B\, V'(N_A) - P_B(N_B - N_A)\frac{d}{dN} V'(N_A)$$

In exactly the same manner,

$$\Delta H = hP_B\, K(N_B) - hP_B\, K(N_A) - hP_B(N_B - N_A)\frac{d}{dN} K(N_A)$$

Substituting for $V' = (2N - 1)\ln R$, K from Eq. 2, and

$$\frac{dV'(N)}{dN} = 2\ln R + \frac{2N - 1}{N(1 - N)}$$

$$\frac{dK(N)}{dN} = \ln R + \frac{2N - 1}{1 - N}$$

gives

$$\Delta U' = P_B\left[(2N_B - 1)\ln\frac{R_B}{R_A} + \frac{(N_B - N_A)(1 - 2N_A)}{N_A(1 - N_A)}\right] = P_B\, F_1(N_B, N_B, N_A) \qquad (3)$$

and

$$\Delta H = hP_B\left[N_B \ln\frac{R_B}{R_A} + \frac{(N_A - N_B)}{(1 - N_A)}\right] \qquad (4)$$

Then the equation for the net desired material holdup is

$$\text{N.D.M.H.} = \frac{2}{\epsilon^2}\,\Delta H - \frac{2}{\epsilon^2}\,(hN_0)\,\Delta U'$$

If we define a general function

$$F_2(N_1,N_2,N_3,N_4) = N_1 \ln \frac{R_2}{R_3} + (N_1 - 1)(R_2 - R_3) - N_4 F_1(N_1,N_2,N_3) \quad (5)$$

where the function F_1 is

$$F_1(N_1,N_2,N_3) = (2N_1 - 1) \ln \frac{R_2}{R_3} + (N_1 - 1)(R_2 - R_3) + \frac{N_1(R_2 - R_3)}{R_2 R_3}$$

Then in a similar fashion it is found that

$$(N.D.M.H.)_A = \frac{2h}{\epsilon_A^2} P_C \, F_2(N_C,N_A,N_0,N_0)$$

$$(N.D.M.H.)_B = \frac{2h}{\epsilon_A^2} P_B \, F_2(N_B,N_B,N_A,N_0)$$

$$(N.D.M.H.)_C = \frac{2h}{\epsilon_C^2} P_C \, F_2(N_C,N_C,N_B,N_0)$$

$$(N.D.M.H.)_{\bar{A}} = \frac{2h}{\epsilon_A^2} P_{\bar{A}} \, F_2(N_A,N_A,N_0,N_0)$$

$$(N.D.M.H.)_{\bar{C}} = \frac{2h}{\epsilon_C^2} P_{\bar{C}} \, F_2(N_A,N_A,N_B,N_0)$$

REFERENCES

Lefkowitz, H., B. Schwartz, and H. Mayer, Columbia Ser. No. 4R-42, Apr. 12, 1943.
Mayer, H., I. Kaplan, and K. Cohen, Report A-396, Dec. 2, 1942.

Appendix D

RAYLEIGH DISTILLATION

The theory of cascades which was presented in Chaps. 1 to 5 was built upon elements that separate according to the law

$$R' = \alpha R \qquad (\alpha = \text{constant})$$
$$R'' = \frac{1}{\beta} R \qquad (\beta = \text{constant}) \tag{1}$$

Another common type of element which does not obey this simple law will now be considered. An example of such an element was first encountered by Lord Rayleigh while he was investigating batch distillation.

Consider the course of a distillation. At any instant the vapor (mole fraction n) is in equilibrium with the composition of the liquid remaining in the flask (mole fraction N″) according to the relation

$$r = \alpha_0 R'' \tag{2}$$

Since the composition of the vapor leaving the liquid is different from that of the liquid, the composition of the residue changes as the distillation proceeds. This is expressed by the conservation relation

$$d(N''Q'') = n\, dQ'' \tag{3}$$

where Q″ is the quantity (in moles) of the liquid residue at any instant.

Combining Eqs. 2 and 3,

$$(\alpha_0 - 1) \frac{dQ''}{Q''} = (1 + \alpha_0 R'') \frac{dN''}{N''}$$

which integrates to give

$$(\alpha_0 - 1) \ln \frac{Q''}{Q} = \ln \frac{R''}{R} + (\alpha_0 - 1) \ln \frac{1 - N}{1 - N''} \tag{4}$$

Here Q is the initial quantity of the charge in the distilling flask, and N is its mole fraction. Introducing the cut $\theta = Q'/Q = (Q - Q'')/Q$, where Q' is the volume of the distillate,

$$\frac{R''}{R} \left(\frac{1 - N}{1 - N''} \right)^{\alpha_0 - 1} = (1 - \theta)^{\alpha_0 - 1} \tag{5}$$

The mole fraction of the destillate, N', is given by

$$N = \theta N' + (1 - \theta) N'' \tag{6}$$

It is evident that Eq. 5 is fundamentally different from Eq. 1 and represents a new kind of separating element.

However if $\alpha_0 - 1$ is small, then

$$\begin{aligned} \frac{R''}{R} &= 1 + (\alpha_0 - 1) \ln (1 - \theta) + \text{order of } (\alpha_0 - 1)^2 \\ \frac{R'}{R} &= 1 + (\alpha_0 - 1) \frac{1 - \theta}{\theta} \ln \frac{1}{1 - \theta} + \text{order of } (\alpha_0 - 1)^2 \end{aligned} \tag{7}$$

Provided θ is constant or nearly so,[1] Eq. 7 is approximately of the form of Eq. 1. Thus the infinitesimal case of Rayleigh distillation can be treated according to the theory of Chaps. 1 to 5, with effective simple-process factor $\alpha = 1 + \epsilon$ given by

$$\begin{aligned} \alpha &= 1 + (\alpha_0 - 1) \frac{1 - \theta}{\theta} \ln \frac{1}{1 - \theta} \\ \epsilon &= (\alpha_0 - 1) \frac{1 - \theta}{\theta} \ln \frac{1}{1 - \theta} \end{aligned} \tag{8}$$

The change in concentration of the residue (N'') and distillate (N') relative to the initial concentration has characteristic peculiarities. From Eq. 7,

[1] It was seen in Eq. 1.19 that it is so in an ideal infinitesimal cascade.

$$N' - N = (\alpha_0 - 1)N(1 - N)\,\frac{1-\theta}{\theta}\ln\frac{1}{1-\theta}$$

$$N'' - N = -(\alpha_0 - 1)N(1 - N)\ln\frac{1}{1-\theta} \qquad (9)$$

from which Table 1 is constructed.

Table 1—Separation and Separative Power in Rayleigh Distillation

θ	$\frac{N' - N}{(\alpha_0 - 1)N(1 - N)}$	$\frac{N'' - N}{(\alpha_0 - 1)N(1 - N)}$	$\frac{2\delta U}{G(\alpha_0 - 1)^2}$	$\frac{2\delta U}{\theta G(\alpha_0 - 1)^2}$
0.0	1.000	−0.000	0.000	1.000
0.1	0.949	−0.1054	0.0999	0.999
0.2	0.892	−0.223	0.1992	0.996
0.3	0.832	−0.357	0.2968	0.990
0.4	0.766	−0.511	0.3914	0.979
0.5	0.693	−0.693	0.480	0.961
0.6	0.611	−0.916	0.559	0.932
0.7	0.516	−1.204	0.622	0.888
0.8	0.402	−1.609	0.647	0.809
0.9	0.256	−2.303	0.589	0.655
0.95	0.158	−2.996	0.473	0.497
1.0	0.000	$-\infty$	0.000	0.000

Key

θ = cut; fraction of charge in distillate
N = mole fraction of charge
N′ = mole fraction of distillate
N″ = mole fraction of residue
α_0 = equilibrium simple-process factor
δU = separative power
G = rate of feed to unit

By taking larger cuts it is possible to increase indefinitely the change in concentration of the residue (column 3 of Table 1). However, the separative power remains finite. If it is supposed that the element continually separates G moles/sec of charge into θG moles/sec of distillate and $(1 - \theta)$G moles/sec of residue, then by Eq. 1.42,

$$\delta U = \frac{G}{2}(\alpha_0 - 1)^2\,\frac{1-\theta}{\theta}\,[\ln(1-\theta)]^2 \qquad (10)$$

The value of δU is found in column 4.

More fundamental than δU is $\delta U/\theta G$, which gives the separative power per unit amount distilled (column 5). It is seen that this term is a maximum $[= \frac{1}{2}(\alpha_0 - 1)^2]$ for $\theta = 0$, decreasing steadily because of mixing of the successive distillate fractions (which are of changing mole fraction). The value of δU itself (column 4) increases initially as θ increases, in spite of the mixing effect, just because more is distilled: as θ approaches 1 the mixing is so serious that it causes a decrease in δU in spite of the increase in amount distilled. (If everything is distilled, evidently no separation is accomplished.) The maximum separative power per unit distillate may be obtained for the entire charge by dividing the distillate into many small portions, instead of collecting it all together. Then the separative power developed reaches $G(\alpha_0 - 1)^2/2$.

It is perfectly plain that any separation process that obeys Eqs. 2 and 3 is formally equivalent to Rayleigh distillation. Of interest is the evaporative centrifuge (Chap. 6, Sec. 4), where the process is indeed a distillation, but the concentration change (Eq. 2) is caused by the centrifugal field instead of by a difference in vapor pressures.

REFERENCES

Cohen, K., Report A-50, Jan. 30, 1941; Report A-60, October 1941.
Cohen, K., and I. Kaplan, Report A-163, Apr. 25, 1942.
Rayleigh, J. W. S., Phil. Mag., 42: 493 (1896).

Appendix E

PROPERTIES OF CONCURRENT TWO-PHASE ELEMENTS

A type of element which occurs in the case of concurrent contacting of two separate phases, and in many other instances, e.g., the concurrent centrifuge, is illustrated in Fig. 1. Figure 2 shows a simple cascade of such units. The element has entrances for two separate

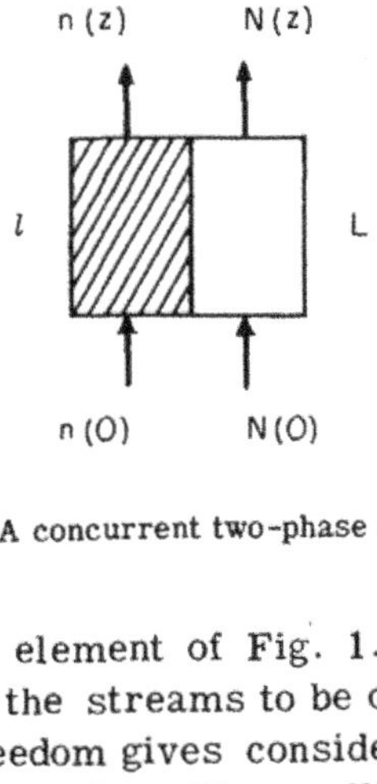

Fig. 1—A concurrent two-phase element.

streams, and unlike the element of Fig. 1.1 (Chap. 1), there is no particular necessity for the streams to be of the same composition. This extra degree of freedom gives considerable flexibility and results in more efficient cascades. There will be exhibited phenomena associated with the degree of equilibrium attained in each unit, and the dependence on cut characteristic of Rayleigh distillation elements (Appendix D) is absent.

The equations governing the transfer of material from one phase to another are

$$L\frac{dN}{dz} = -k\,[(N-n) - (\alpha-1)N(1-N)]$$

$$l\frac{dn}{dz} = +k\,[(N-n) - (\alpha-1)N(1-N)] \qquad (1)$$

provided

$$\frac{N}{1-N} = \alpha \frac{n}{1-n} \qquad \alpha > 1$$

in the equilibrium state. Here $k > 0$ is a transfer coefficient and z is a parameter expressing the size of the unit. For example, in a catalytic-exchange process, z would be the volume of catalyst and k would be proportional to the surface area per unit volume of catalyst.

These equations are consistent with the equations of conservation

$$L\,N(z) + l\,n(z) = \text{constant} = L\,N(0) + l\,n(0) \tag{2}$$

and solving Eqs. 1 and 2 together it is found that, after some rearranging,

$$\begin{aligned} N(z) - N(0) &= (1-\theta)\,[n(0) - N(0) + (\alpha - 1)N(1-N)]\,(1 - e^{-bz}) \\ n(z) - n(0) &= -\theta\,[n(0) - N(0) + (\alpha - 1)N(1-N)]\,(1 - e^{-bz}) \end{aligned} \tag{3}$$

where

$$\begin{aligned} \theta &= \frac{L}{L + l} \\ b &= \frac{k}{L + l}\,\frac{1}{\theta(1-\theta)} \end{aligned} \tag{4}$$

When $z \to \infty$, the last term in Eq. 3 approaches 1, so that

$$\beta \equiv 1 - e^{-bz} \tag{5}$$

is the percentage of equilibrium attained in the unit.

The separative power of the element is

$$\delta U = L\,V[N(z)] - L\,V[N(0)] + l\,V[n(z)] - l\,V[n(0)] \tag{6}$$

Expanding the value functions about $V[N(0)]$ and collecting,

$$\delta U = L[N(z) - N(0)]\,V''[N(0)]\left\{\frac{1}{2}[N(z) - N(0)] + N(0) - n(0) + \frac{1}{2}[n(0) - n(z)]\right\}$$

and substituting from Eq. 3,

$$\delta U = \frac{L(1-\theta)}{2N^2(1-N)^2}\beta(2-\beta)\left\{(\alpha-1)N(1-N) - [N(0)-n(0)]\right\}\left[\frac{\beta}{2-\beta}(\alpha-1)N(1-N) + N(0) - n(0)\right] \quad (7)$$

It is seen that δU is a function of β and $N(0) - n(0)$. Taking $N(0) = n(0)$, which corresponds to the no-mixing case for single-entrance elements, the following is obtained:

$$\delta U = \frac{(L + l)\theta(1-\theta)(\alpha-1)^2}{2}\beta^2$$

and from Eq. 4, the separative power per unit length is

$$\frac{\delta U}{z} = \frac{k(\alpha-1)^2}{2}\,\frac{\beta^2}{-\ln(1-\beta)} \qquad \text{(no-mixing case)} \qquad (8)$$

which has a maximum at $\beta = 0.715$.

$$\frac{\delta U}{z} = 0.40724\,\frac{k(\alpha-1)^2}{2} \qquad \text{(no-mixing case)} \qquad (9)$$

However, this choice of $N(0) - n(0)$ is not the most efficient. Differentiating δU from Eq. 7 with respect to $N(0) - n(0)$, it is found to be a maximum when

$$N(0) - n(0) = \frac{1-\beta}{2-\beta}(\alpha-1)N(1-N) \qquad (10)$$

which is

$$\frac{\delta U}{z} = \frac{1-\beta}{(2-\beta)\ln(1-\beta)}\,\frac{k(\alpha-1)^2}{2} \qquad (11)$$

and a relative maximum with respect to β when $\beta = 0$, namely,

$$\frac{\delta U}{z} = \frac{k(\alpha-1)^2}{4} \qquad \text{(optimum concentration relation)} \qquad (12)$$

By comparing Eq. 12 with Eq. 9 it is seen that the optimum concentration relation permits an increase in separative power of 19.6 per cent over the no-mixing case.

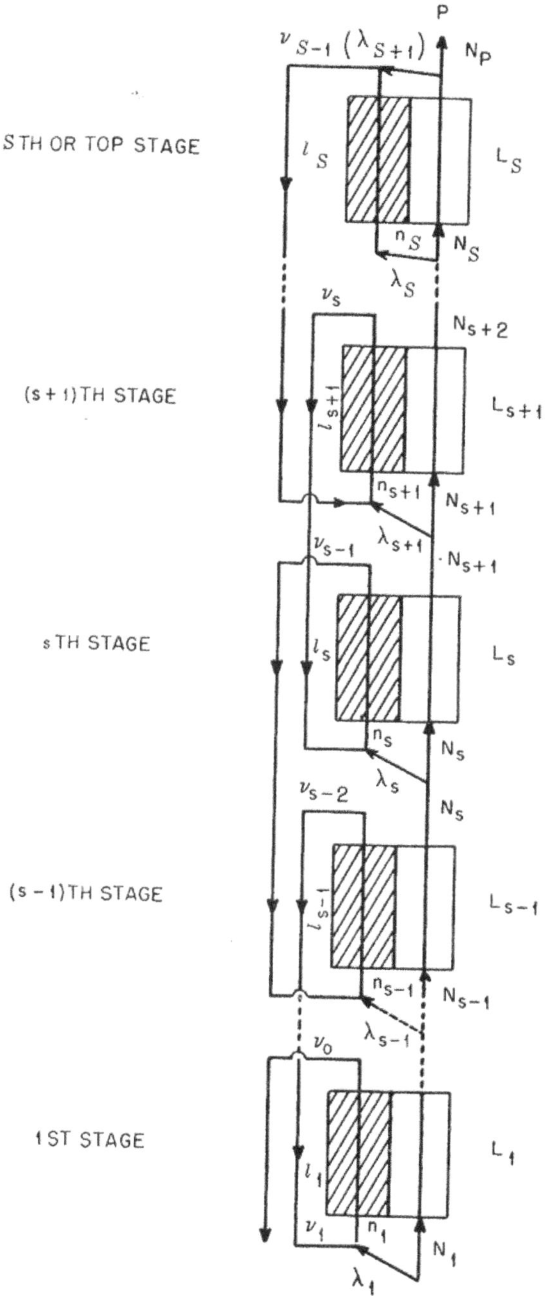

Fig. 2—A cascade of concurrent double-entrant units (rectifying section).

It is further evident from Eqs. 4, 8, and 11 that $\theta = \frac{1}{2}$ is the most efficient cut, because this value results in a minimum value of $(L + l)$ for a given β.

It is instructive to compare these two cases (no-mixing and optimum-concentration relation) by considering the net transport for the cascade of Fig. 2 with $\theta = \frac{1}{2}$. This is

$$L_s(N_{s+1} - \nu_s) = P(N_p - \nu_s)$$

On the other hand, for a given flow L_s the maximum possible transport through the *s*th stage is

$$L_s(\alpha - 1)N_{s+1}(1 - N_{s+1})$$

which occurs when

$$\frac{\nu_s}{1 - \nu_s} = \frac{1}{\alpha}\frac{N_{s+1}}{1 - N_{s+1}}$$

The percentage transport (referred to the maximum) is therefore

$$q = \frac{N_{s+1} - \nu_s}{N_{s+1}(1 - N_{s+1})(\alpha - 1)} = \frac{N(0) - n(z)}{(\alpha - 1)N(1 - N)} \tag{13}$$

When $N(0) - n(0)$ are related by Eq. 10 and $\theta = \frac{1}{2}$, q turns out to be $\frac{1}{2}$; when they are equal, $q = \beta/2$. Thus in the no-mixing case the net transport is related to the degree of equilibrium.

The results just obtained may be generalized by introducing the parameter q directly into the formula (Eq. 7) for δU. It is then found that (for $\theta = \frac{1}{2}$)

$$\frac{\delta U}{z} = -2k(\alpha - 1)^2 \frac{\beta q(1 - q)}{(2 - \beta)\ln(1 - \beta)} \tag{14}$$

$$N(z) - N(0) = \frac{(1 - q)\beta}{2 - \beta}(\alpha - 1)N(1 - N) \tag{15}$$

From Eqs. 14 and 15 expressions are obtained for the total volume of a cascade, the total pumping, and the total number of stages. These are, for the constant-transport case,

$$\text{Total volume} \approx -\frac{(2-\beta)\ln(1-\beta)}{q(1-q)\beta}$$

$$\text{Total pumping} \approx \frac{1}{q(1-q)}\,\frac{2-\beta}{\beta}$$

$$\text{Number of stages} \approx \frac{2-\beta}{(1-q)\beta}$$

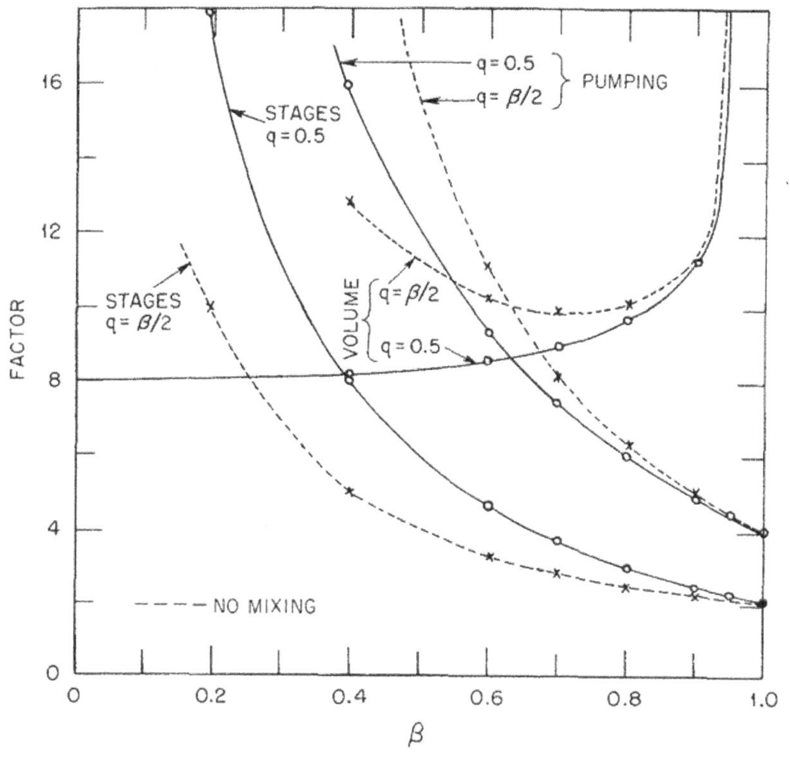

Fig. 3—Factors for total pumping, volume, and number of stages; a comparison of no-mixing cascade and "constant per cent transport" cascade (q = 0.5).

The volume and pumping are both minimum with respect to q for q = ½. Figure 3 gives the three factors as a function of β for the two choices q = 0.5 (minimum volume) and q = $\beta/2$ (no mixing). The minimum volume for q = 0.5 is at β = 0 and is 19 per cent less than the minimum volume for the no-mixing case which occurs at $\beta = 0.715$. Unfortunately both the pumping and the number of stages are infinite

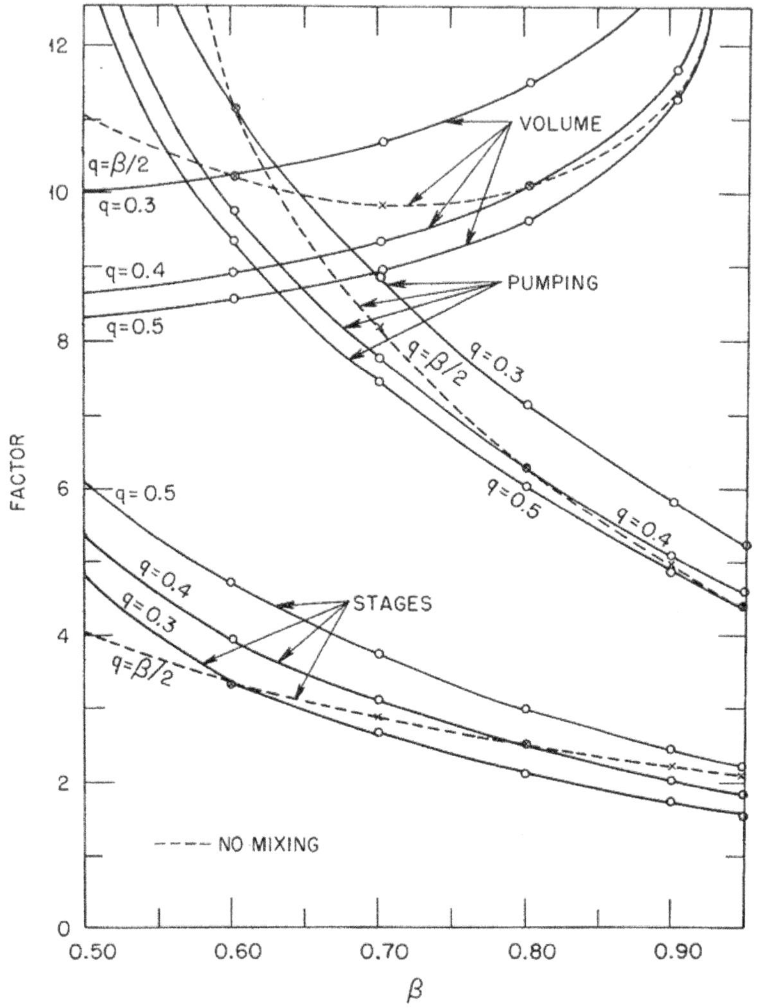

Fig. 4—Factors for total pumping, volume, and number of stages as functions of q and β.

at $\beta = 0$. It is seen, therefore, that the significant region will lie between $\beta = 0.95$ and $\beta = 0.50$. Figure 4 gives the factors for volume, pumping, and number of stages for $q = 0.3$, 0.4, 0.5, and $\beta/2$ within this range of β.

It is obvious that the more general "constant per cent transport" cascade offers greater opportunities than does the no-mixing case which might a priori be considered the most efficient. A few interesting possibilities are noted which illustrate the flexibility of the constant-transport cascades.

The factor for minimum volume in the no-mixing case is 9.82 and corresponds to a 71.5 per cent approach to equilibrium. In the constant-transport case this volume factor corresponds to many choices of q and β; among others, to $q = 0.5$ and $\beta = 0.82$. In this case, however, the total pumping is 24 per cent less than in the no-mixing case, whereas the number of stages is only 3 per cent higher. It is thus possible to save a considerable amount of pumping while losing only slightly on the number of stages. Another constant-transport possibility with the volume 9.82 is the choice $q = 0.4$ and $\beta = 0.77$. The pumping is 15 per cent less and the number of stages 4 per cent less than the no-mixing values for $\beta = 0.715$. The no-mixing case thus represents no extremal.

A good balance between volume, pumping, and stages will be struck by the choice $q = 0.40$ and $\beta = 0.80$. But if one of these three factors overweighs in importance the others, different values should be chosen. Thus if it is especially desirable to reduce the volume to a minimum, the choice might be $q = 0.5$ and $\beta = 0.5$. If it is desirable to reduce to the limit both the pumping and the number of stages, the choice should be $q = 0.4$ and $\beta = 0.90$, or even $\beta = 0.95$. If the number of stages is not important, then $q = 0.5$ would be chosen.

REFERENCES

Cohen, K., and I. Kaplan, Report A-302, Sept. 19, 1942.
Kaplan, I., and K. Cohen, Report A-301, Oct. 29, 1942.

INDEX

A

Aston, F. W., 103

B

Bardeen, J., 131
Beams, J. W., 105, 113
Boundary conditions, rectifier, 87
  stripper, 87-88
Bramley, A., 105, 119, 124, 131
Brewer, A. K., 105

C

Campbell, G. A., 50
Cascade, 3, 5
  combined, 78
  cut in, 5
  design of, 11
  efficiency of, 6
  equilibrium time of, 24-27, 38, 82
  flow in, 13-15
    total, 9, 77-78
  ideal, 5-10
    stages of, 5
  ideal and nonideal, 10-12
    losses in, 67-74
      per stage, 69
  production rate of, 12, 94-102
  rectifier of, 21
  relaxation time of, 25
  square, 30-44
    equilibrium time of, 39, 144-146
    production loss of, 101
    production rate of, 94, 96
      disturbance function of, 96
    steady state of, 31, 62, 65
    transient phenomena of, 62-64
    Type I operation of, 40-43, 47, 54
Cascade, square, Type II operation of, 41-44, 47-53
  squared-off, of countercurrent elements, 34-35
    rapid approximate solution for, 35-38
    scheme for, 38
  stages of, 3-5
    number of, 3, 12
    size of, 9
  stripping sections of, 21-24, 77-78
Centrifuge, concurrent flow, 105, 114
  countercurrent flow, 105, 119, 121-122
  evaporative, 104-105, 111-114
  separative power of, 109-111
Centrifuge separation, design of, 105-106
  differential equation of, 106, 109
  history of, 103-106
  maximum separative power of, 109-111
  theory of, 105
Chapman, S., 131
Cohen, K., 29, 38, 61, 78, 124-125, 136, 143, 149, 153, 161
Column, countercurrent, 31
    number of stages, 31-32
  fractionating, 30
  relaxation phenomena for, 54-61
  thermal diffusion, 30
Combination of separation plants, 74-78
Concurrent contacting, 131
Concurrent two-phase elements, 154-161
Control, hydrodynamic, 72-102
Control problem, 79-102
Countercurrent centrifuge, 30
Countercurrent column, 31
  number of stages, 31-32
Countercurrent contacting, 129
Countercurrent devices, 30-31
Cowling, T. G., 131
Cut in cascade, 5

D

Debye, P., 131
Deuterium, separation of, 126, 132
Deuterium separation, catalytic exchange
  method, 132-136
Diffusion, thermal, 131-132
Dirac, P. A. M., 29, 109, 125
Distillation, Rayleigh, 150-153
  separation in, 152
  separative power in, 152
Doetsch, G., 94
Dunning, J. R., 29, 38

E

Elements, number of, 14, 19
Equilibrium fractionation, 67
Equilibrium time, 24-27, 38, 82
  of square cascades, 39, 144-146

F

Feynman, R. P., 67
Flow, total, 9, 15, 23
Fluctuations, 80-90
  nonperiodic, 90
  periodic, 90
Foster, R. M., 50
Fuchs, K., 29, 84, 91, 102
Furry, H., 119, 131

G

Gevrey, M., 145

H

Holdup function, 25, 77-78, 147-149
  in column, 59
  net desired material (N.D.M.H.),
    25-27
  in top reservoir, 59
Humphreys, R. F., 105, 113
Hutchison, C. A., 61, 136

I

Isotope separation, 27-29
  by chemical exchange, 126-127
  by thermal diffusion, 131
  uranium, 15

J

Joly, J., 104
Jones, R. C., 119, 131

K

Kaplan, I., 29, 38, 61, 78, 106, 125, 136,
  143, 149, 161
Keith, P. C., 136
Kuhn, W., 105, 119
Kurti, N., 29

L

Laplace transform, 84, 89, 93
Lefkowitz, H., 78, 149
Lindemann, F. A., 103

M

Martin, H., 105, 119
Mayer, H., 78, 84, 149, 161
Mixing, 6, 15
Mole fraction, 5
Mulliken, R. S., 104
Murphree, E. V., 125

N

Net transport, 21, 26

O

Onsager, L., 119, 131

P

Peierls, R., 29
Plants, separation, combination
  of, 74-78
Poole, J. H. J., 104
Process factor, centrifuge, 105
  infinitesimal, 9
  large, 15
  simple, 5
Production rate, variations in, 94-102

R

Rayleigh, J. W. S., 150, 153
Relaxation phenomena, 54-61

S

Schwartz, B., 78, 149
Separation, isotope (see Isotope separation)
  two-phase, 126-131
Separation plants, combination of, 74-78
Separative power, 13-15, 19-21, 37-38
  (See also Centrifuge; Distillation)
Separative work, 72
Simon, F., 29
Skarstrom, C., 105-106, 113
Square cascade (see Cascade, square)
Stages of cascade (see Cascade)

T

Thermal diffusion, 131-132
Thompson, W. I., 62, 78, 136
Transcendental equation, roots of, 139- 143
Transport, loss of, 90-91
  maximum, 33

U

Urey, H. C., 105-106, 125, 133

V

Value function, 13-14
  elementary, 15, 17

W

Waldman, L., 131
Waste material, 18
Wirtz, K., 131
Work function, 64

Made in the USA
Monee, IL
05 November 2026

67815677R00105